Communications in Computer and Information Science   1107

Commenced Publication in 2007
Founding and Former Series Editors:
Phoebe Chen, Alfredo Cuzzocrea, Xiaoyong Du, Orhun Kara, Ting Liu,
Krishna M. Sivalingam, Dominik Ślęzak, Takashi Washio, Xiaokang Yang,
and Junsong Yuan

Editorial Board Members

More information about this series at http://www.springer.com/series/7899

Luca Longo · Maria Chiara Leva (Eds.)

Human Mental Workload

Models and Applications

Third International Symposium, H-WORKLOAD 2019
Rome, Italy, November 14–15, 2019
Proceedings

 Springer

Editors
Luca Longo ⓘD
Technological University Dublin
Dublin, Ireland

Maria Chiara Leva ⓘD
Technological University Dublin
Dublin, Ireland

ISSN 1865-0929 ISSN 1865-0937 (electronic)
Communications in Computer and Information Science
ISBN 978-3-030-32422-3 ISBN 978-3-030-32423-0 (eBook)
https://doi.org/10.1007/978-3-030-32423-0

This Springer imprint is published by the registered company Springer Nature Switzerland AG
The registered company address is: Gewerbestrasse 11, 6330 Cham, Switzerland

Preface

This book endeavors to stimulate and encourage discussion on mental workload, its measures, dimensions, models, applications, and consequences. It is a topic that demands a multidisciplinary approach, spanning across human factors, computer science, psychology, neuroscience, statistics, and cognitive sciences. This book presents recent developments in the context of theoretical models of mental workload and practical applications.

This year in particular it contains a selection of the work presented in the context of the Third International Symposium on Mental Workload, Models and Applications (H-WORKLOAD 2019), sponsored by the Sapienza University of Rome and supported by the Irish Ergonomics Society. It contains a revision of the best papers presented at the symposium and selected through a strict peer-review process. The contributions of this edition were predominately focused on the use of neurosciences tools in the context of detecting, assessing, and modeling mental workload.

From the content of these research contributions, it is clear that mental workload, as a multidimensional and multifaceted construct, is still under definition, development, and investigation. This is one of the reasons why mental workload is today a keyword used and abused in life sciences, as pointed by Prof. Fabio Babiloni. However, despite the difficulty in precisely defining and modeling it, the capacity to assess human mental workload is a key element in designing and implementing information-based procedures and interactive technologies that maximize human performance. Some of the articles published in this book applied psychological subjective self-reporting measures, others made use of primary task measures and some a combination of these. Physiological measures in general, and more specifically electroencephalography (EEG), have been gaining a more prominent role, thanks to advances in data-gathering technology as well as a growing availability of computational power and classification techniques offered by the discipline of artificial intelligence. This is also reflected in the present book where half of the chapters focus on the development of novel models of mental workload employing data-driven techniques, borrowed from machine learning. However, one of the key issues in modeling mental workload employing automated learning techniques is that, although it often leads to accurate and robust models, they lack explanatory capacity. This problem is fundamental if we want to define mental workload for the fields of human factors, human–computer interaction, and in general for human-centered designers. Thus, we believe that future research efforts on mental workload modeling should employ a mix of measures as well as qualitative and quantitative research methods to not only assess mental workload but also to understand its meaning and implications on the individuals and our approach toward work and life.

September 2019

Luca Longo
M. Chiara Leva

Preface

September 2019

Acknowledgments

We wish to thank all the people who helped in the Organizing Committee for the Third International Symposium on Mental Workload, Models and Applications (H-WORKLOAD 2019). In particular the local chairs, Dr. Gianluca Di Flumeri, Dr. Gianluca Borghini, Dr. Pietro Aricò, and the members of the scientific committee listed in this book for their contribution toward the revision process and their constructive feedback to the scientific community around it. We would also like want to include in our thanks the main sponsors of the event, the Sapienza University of Rome, the Irish Ergonomics Society, and the Technological University Dublin. Without the support of this extended community the conference and the book would not have been realized. A special thank you goes to the researchers and practitioners who submitted their work and committed to attend the event and turn it into an opportunity to meet and share our experiences on this fascinating topic.

Organization

Organizing Committee

General Chairs, Editors, and Program Chairs

Luca Longo   Technological University Dublin, Ireland
Maria Chiara Leva   Technological University Dublin, Ireland

Local Chairs

Gianluca Di Flumeri   Sapienza University of Rome, Italy
Gianluca Borghini   Sapienza University of Rome, Italy
Pietro Aricò   Sapienza University of Rome, Italy

Program Committee

Andy A. Smith   Psychology Cardiff University, UK
Mobyed Ahmed   Mälardalen University, Sweden
Julie Albentosa   French Armed Forces Biomedical Research Institute,
                              France
Pietro Aricò   Sapienza University of Rome, Italy
Bargiotas   Paris Saclay Paris, France
Emanuele Bellini   Khalifa University, UAE
Gianluca Borghini   Sapienza University of Rome, Italy
Bethany Bracken   Charles River Analytics, USA
Suzy Broadbent   BAE Systems, UK
Karel Brookhuis   The University of Groningen, The Netherlands
Aidan Byrne   Swansea University, UK
Brad Cain   CAE Defence and Security, Canada
Jose Cañas   University of Granada, Spain
Giulia Cartocci   Sapienza University of Rome, Italy
Martin Castor   Group of Effectiveness, Interaction, Simulation,
                              Technology and Training, Sweden
Loredana Cerrato   Nuance Communications, USA
Lorenzo Comberti   Politecnico di Torino, Italy
Ferdinand Coster   Yokogawa Europe Solutions B.V., UK
Katie Crowley   Trinity College Dublin, Ireland
Dick De Waard   The University of Groningen, The Netherlands
Micaela Demichela   Politecnico di Torino, Italy
Gianluca Diflumeri   Sapienza University of Rome, Italy
Tamsyn Edwards   San Jose State University, NASA AMES, USA
Enda Fallon   NUI Galway, Ireland
Jialin Fan   Cardiff University, UK

Contents

About the Editors

Dr. Luca Longo is currently an assistant professor at the Technological University Dublin, where he is a member of the Applied Intelligence Research Centre. His core theoretical research interests are in artificial intelligence, specifically in automated reasoning and machine learning. He also performs applied research in mental workload modeling. He is the author of 50+ peer-reviewed articles that have appeared in conference proceedings, book chapters, and journals in various theoretical and applied computer science fields. Luca has been awarded the National Teaching Hero in 2016, by the National Forum for the Enhancement of Teaching and Learning in Higher Education.

Dr. Maria Chiara Leva is a lecturer at the Dublin Institute of Technology in the College of Environmental Health. She is also a visiting research fellow in the Centre for Innovative Human Systems at Trinity College Dublin. Her area of expertise is human factors and safety management systems. Chiara holds a PhD in Human Factors conferred by the Polytechnic of Milan, Department of Industrial Engineering. Her PhD focused on human and organizational factors in safety-critical system in the transport sector. She is the co-chair of the Human Factors and Human Reliability Committee for the European Safety and Reliability Association and has been working in ergonomics and risk assessment as a consultant since 2008. She has authored more than 56 publications on human factors, operational risk assessment, and safety management in science and engineering journals.

Models

Mental Workload Monitoring: New Perspectives from Neuroscience

Fabio Babiloni[1,2(✉)]

[1] Department of Molecular Medicine, Sapienza University of Rome, Rome, Italy
fabio.babiloni@uniromal.it
[2] Department of Computer Science, Hangzhou Dianzi University,
Hangzhou, China

Abstract. Mental Workload is nowadays a keyword used and sometimes abused in life sciences. The present chapter aims at introducing the concept of mental workload, its relevance for Human Factor research and the current needs of applied disciplines in a clear and effective way. This paper will present a state-of-art overview of recent outcomes produced by neuroscientific research to highlight current trends in this field. The present paper will offer an overview of and some examples of what neuroscience has to offer to mental workload-related research.

Keywords: Mental workload · Operational environments · Neuroscience · EEG · Neurometrics · BCI

1 Introduction to the Mental Workload Concept

Mental Workload is a complex construct that is assumed to be reflective of an individual's level of cognitive engagement and effort while performing one or more tasks [1]. Therefore, the assessment of mental workload can essentially be a quantification of mental activity resulting from such tasks. It is difficult to give a unique definition of *Mental Workload*. Various definitions have been given during the last decades:

- "Mental workload refers to the portion of operator information processing capacity or resources that is actually required to meet system demands" [2];
- "Workload is not an inherent property, but rather it emerges from the interaction between the requirements of a task, the circumstances under which it is performed, and the skills, behaviours, and perceptions of the operator" [3];
- "Mental workload is a hypothetical construct that describes the extent to which the cognitive resources required to perform a task have been actively engaged by the operator" [4];
- "The reasons to specify and evaluate the mental workload is to quantify the mental cost involved during task performance in order to predict operator and system performance" [5].

Apart from the definitions presented above, many other attempts to uniquely define Mental Workload concept have been made, demonstrating how mental it may not be a unitary concept because it is the result of different aspects interacting with each other.

L. Longo and M. C. Leva (Eds.): H-WORKLOAD 2019, CCIS 1107, pp. 3–19, 2019.
https://doi.org/10.1007/978-3-030-32423-0_1

In fact, several mental processes such as alertness, vigilance, mental effort, attention, stress, mental fatigue, drowsiness and so on, can be involved in task execution and they could be affected by specific tasks demand in each moment. In general, mental workload theory assumes that: (i) people have a limited cognitive and attentional capacity, (ii) different tasks will require different amounts (and perhaps different types) of processing resources, and (iii) two individuals might be able to perform a given task equally well, but differently in terms of brain activations [6, 7].

1.1 A Topic in the Human Factors Research

In some safety-critical operational environments, one or few operators could be responsible of the safety, and even more the life, of numerous people. For example, let us think to the transportation domain (e.g. Aviation, Rail, Maritime), where the safety of the passengers depends on the performance of the Pilot/Driver/Sailor, the Traffic-Controller or the Maintenance crew. In such contexts, a human error could have serious and dramatic consequences. In general, human error has consistently been identified as one of the main factors in a high proportion of all workplaces accidents. In particular, it has been estimated that up to 90% of accidents exhibits human errors as principal cause [8]. This is true also for other domains such as health care, the US Institute of Medicine estimates that there is a high people mortality per year (between 44.000 and 88.000) as a result of medical errors [9], with an impressive amount of accidents resulting from breast cancer treatments that doubles fatalities resulting from road accidents [10, 11]. Scientific publications regarding problem of medical surgeries, injuries and complications of treatment can be fairly dated to the 1991 as results of the Harvard Medical Practice Study [10, 11]. The reviews of 30,000 medical records of patients hospitalized in the New York state showed that the 4% of patients had complications of their treatment, which have been called *Adverse Events* (AE). Even more shocking was the finding that two-thirds of these injuries were due to medical operators' mistakes, highlighting the fact that they were preventable. The US study was replicated in other Countries [12, 13] with the same results trend (Australia: 13% of patients with *AE*; UK: 10%). The report *"To Err is Human"* of the *Institute of Medicine* (IOM), published in the 2000, had a dramatic effect in bringing patient safety to the medical and public attention. The IOM proclaimed that nationwide as many as 98,000 Americans died yearly because of medical mistakes [14]. It has also been estimated that inappropriate human actions and consequently the errors implicated are the main causes of 57% of road accidents and a contributing factor in over 90% of them [15]. The Aviation Safety Network reported 19 accidents with 960 casualties during the last years; in many cases factors related to workload, situation awareness and monitoring were a caused or contributing factors [16, 17]. Additionally, over the past four decades, human error has been involved in a high number of casualty catastrophes, including the Three Mile Island, Chernobyl and Bhopal nuclear power disasters, the Tenerife, Mont St Odile, Papa India and Kegworth air disasters, the Herald of Free Enterprise ferry disaster, the Kings Cross fire disaster, the Lad-broke Grove rail disaster, and many others [18]. Consequently, the human factor concern, with its possible causes and ways to mitigate its impact, received more and more attention, and it has been investigated across a wide range of domains, including military and civil aviation [19, 20], aviation maintenance

[21], air traffic management [22, 23], rail [24], road transportation [25, 26], nuclear power and petrochemical reprocessing [27], military, medicine [9, 28], and even the space travel domain [29]. At this point, what are the causes of human errors? Human error is an extremely common phenomenon: people, regardless of abilities, skill level and expertise, makes errors every day. The typical consequence of error-occurrence is the failure to achieve the desired outcome or, even worst, the production of an undesirable outcome. When it happens in particular working environments, such error can potentially lead to accidents involving injuries and fatalities. Human error can be defined as the execution of an incorrect or inappropriate action, or a failure to perform a particular action. According to the scientific literature, there have been numerous attempts at de-fining and classifying the human error. However, a universally accepted definition does not yet exist. Rasmussen [30] pointed out the difficulty in providing a satisfactory definition of human error. In 1987, he suggested that "human error represents a mismatch between the demands of an operational system and what the operator does" [31]. The main causes of human errors can be searched within the internal or psychological factors of the operator [32]. In fact, errors could also arise from aberrant mental processes such as inattention, poor motivation, loss of vigilance, mental overload and fatigue that negatively affect the performance.

Among the various cognitive components of mental activity, mental workload is anyway considered to be the one indicating a comprehensive representation of an operator's mental state considering the amount of involved cognitive resources, therefore cognitive psychology aimed to establish the relationship between mental workload and human performance. Modelling such a relationship would help to predict human performance evolution along time thus preventing potentially risky situations. In this sense, the widest accepted hypothesis describes the relationship between mental workload and performance through an "inverted U-shape" (Fig. 1) function. This hypothesis relies on the Yerkes-Dodson theory, that more than one century ago (Robert M. Yerkes and John D. Dodson, 1908) described the effects on human performance referred to as physiological activation [33]. Reasonably, this theory has not to be intended as an exact one, but as a representative model, perhaps revised in different ways recently [34, 35], however the pillar is that such a relationship is not linear, and performance tends to degrades at both the boundaries of workload span. In other words, some levels of mental workload may help the user to reach high performance levels [36] since it stimulates positively the user and it keeps him/her awake with high attention level. Nevertheless, a period of mental inactivity and "under-stimulation" can cause a monotonous and boring state (underload), a low level of vigilance and attention, with low cognitive resources demand. For example, Warm and colleagues [37] showed how vigilance requires an important amount of cognitive resources, by using behavioural, neurophysiological and subjective measures. At the same time, an operative condition characterized by highly demanding multiple tasks can lead the user to an overload condition, equally impairing from a performance perspective [38, 39]. Both the cases bring to a variation in neurophysiological factors and often to a decrement of performance. Such performance reduction is highly undesired, especially in safety-critical domains, as discussed above.

In 1981 Wickens pointed out that the development of increasingly complex technologies was radically changing the role and load to which an operator was subjected,

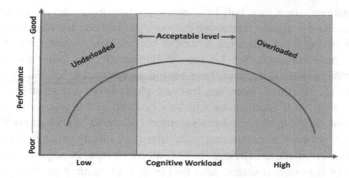

Fig. 1. Inverted U-shape relationship between mental workload and performance.

leading to the dual need to exploit the model of multiple resources to optimize the processing of human operator information in the definition of tasks ("Should one use keyboard or voice? Spoken words, tones, or text? Graphs or digits? Can one ask people to control while engaged in visual search or memory rehearsal?" [40]) and measure the operator's workload [41].

Since then, the measure of the workload has spread from the aeronautical field [16, 42] to the educational [43] and to the clinical [44] domains. Mental Workload assessment techniques should be able to solve these questions, being sensitive to fluctuations in task cognitive demands while operating or interacting with systems by without intruding external interferences on primary task performance [45, 46]. To this regard, three are the main categories of techniques investigated and employed in Ergonomic field for mental workload monitoring [47, 48]:

- Behavioural measures, generally derived from measures of operator's performance with respect to the main and/or an additional task;
- Subjective measures, generally collected through self-confrontation reports and questionnaires, such as the Instantaneous Self-Assessment (ISA, [49]) and NASA-Task Load indeX (NASA-TLX, [3]);
- Neurophysiological measures, i.e. those techniques that infer human mental states from specific variations of human biosignals, such as brain activity, hearth activity, skin sweating, and so on.

The targeted level of sensitivity is unobtainable with behavioural and subjective measures alone. In this regard, neurophysiological techniques have been demonstrated to be able to assess mental workload of humans with a high reliability, even in operational environments [50–54]. Moreover, neurophysiological techniques afford another important advantage: unlike alternative subjective assessment techniques, neurophysiological measures do not require the imposition of an additional task either concurrently (as in secondary task techniques) or subsequently (as in subjective workload assessment techniques) the primary one. Neurophysiological measures can be obtained continuously, even online, with little intrusion, i.e. without interrupting the operator's work with additional tasks or questions [50]. In addition, it will become more and more difficult to measure cognitive capacities with performance indices in future workstations, since they will be characterized by higher levels of automation, therefore reducing the manual interaction between the humans and the machine. Also, any eventual performance degradation would become "measurable" by the system when the operator already suffered a mental impairment, i.e. "after the fact" [55]. Finally, neurophysiological measures have been demonstrated to be reliably diagnostic of multiple levels of arousal, attention, learning phenomena, and mental workload in particular [56–65]. Such applications will be discussed in the following paragraphs. Thus, the online neurophysiological measurements of mental workload could become very important, not only as monitoring techniques, but mainly as support tools to the user during his/her operative activities. In fact, as the changes in cognitive activity can be measured in real-time, it should also be possible to manipulate the task demand (adaptive automations) in order to help the user to keep optimal levels of mental workload under which he or she could be operating [66]. In other words, the neurophysiological workload assessment could be used to realize a *passive Brain Computer Interface* (passive-BCI, please see Par. 3) application in real environments.

2 Neuroscientific Contribute to the Mental Workload Assessment

Neuroimaging methods and cognitive neuroscience have steadily improved their scientific and technological maturity over the past decade, consequently producing a growing interest in their use to examine the neural circuits and physiological phenomena behind human complex tasks representative of perception, cognition, and action as they occur in operative settings. At the same time, many fields in the biological sciences, including neuroscience, are being challenged to demonstrate their relevance to practical real-world problems [48]. Although the different conjugations of the discipline name, for example Bioengineering or Cognitive Neuroscience or Neuroergonomy (in general due to a different point of view of the same problem), scientific research in these fields is aiming at inferring and assessing humans' workload, and more generally their *Internal States (IS)*, through neurophysiological measures. In fact, such a concept of measuring humans' IS, where IS has to be intended as the generalization of all the possible mental, or purely cognitive (e.g. workload, attention, situation awareness), and affective, or purely emotional (e.g. stress, pleasantness, frustration) psychophysical states, is based on the assumption that each biological activity is regulated by the human Central Nervous System (CNS). The brain is of course the main actor of CNS, but it is also important to take into account the activity of the Autonomic Nervous System, that acts largely unconsciously and regulates bodily functions such as the heart rate, respiratory rate, pupillary response, skin sweating and so on [67]. Variations of these biological activities correspond to internal reactions because of modification of external (environment) and internal (cognition, emotions, etc.) factors, therefore neurophysiological signals become the interface to access what is happening within the human mind. Just to make few examples, the electrical activity of the brain's prefrontal cortex in EEG Theta frequency band increases while cognitive demand is increasing [68], increased skin sweating is related to higher levels of arousal and attention [69], while heart rate tends to accelerate under stress conditions [70]. The last decades have been fruitfully spent by the scientific community in investigating the correlation between such variations and specific human ISs, enabling the possibility of obtaining measures (i.e. *Neurometrics*), of several concepts such as attention, stress, workload, emotion, etc., and to use them (i) to provide a feedback to the user [56]; (ii) to modify the behaviour of the interface the user is interacting with [71]; or (iii) to obtain insights related to user's feelings while experiencing specific situations (e.g. eating, or watching tv, or any other everyday activity) without any verbal communication [72, 73]. These potentialities have been usually demonstrated in controlled settings (i.e. Laboratory), but recent technological progresses are allowing more and more low invasive and low cost wearable devices that can open the door to applications in everyday natural settings. From a technological point of view, many companies are moving forward to develop biosignal acquisition devices more and more wearable and minimally invasive and at the same time sensors (e.g. gel-free electrodes for EEG systems, or bracelets/watches already integrating sensors for PPG and GSR) able to ensure high signal quality and comfort at the same time [74, 75]. Just to have an idea of the effort recently produced in this field, many works have been performed in operational environments, e.g. aviation [50, 76–78], surgery [79], city traffic monitoring [80–82], power plant control centres [83], and many others [59, 73, 84–86] to demonstrate the usefulness of Neurometrics.

Such neuroscientific researches are based on the use of neuroimaging technologies and neurophysiological measures, including *Electroencephalography* (EEG), *functional Near-InfraRed* (fNIR) imaging, *functional Magnetic Resonance Imaging* (fMRI), *Magnetoencephalography* (MEG), and other types of biosignals such as *Electrocardiography* (ECG), *Electrooculography* (EOG) and *Galvanic Skin Response* (GSR) [68, 87, 88]. Neuroimaging methods such as *Positron Emission Tomography* (PET) and fMRI are excellent tools in this endeavour, enabling the examination of how the brain adapts itself in response to practice or repeated exposure to particular tasks. However, their limitations in terms of cost, space and invasiveness make them not suitable for real working environment settings, where a less invasive approach would be preferable and the costs for its implementation and usage has to be limited. In fact, PET and fMRI techniques require expensive instruments and high maintenance costs, In addition, fMRI [89] and MEG techniques require room-size equipment that are not portable. On the other hand, EOG, ECG and GSR activity measurements highlighted a correlation with some mental states (stress, mental fatigue, drowsiness), but they were demonstrated to be useful only in combination with other neuroimaging techniques directly linked to the *Central Nervous System* (CNS), i.e. the brain [68, 90, 91]. Consequently, the EEG and fNIRs are the most likely candidates that can be straightforwardly employed to investigate human brain behaviours in operational environments. The propensity for using EEG or fNIRs techniques has not been clarified yet. There are several factors to take into account about real operative scenarios. For example, both EEG and *Fast Optical Signal* (FOS)-based fNIR have similar bandwidth and sample rate requirements, as the FOS appears to directly reflect aggregated neural spike activity in real-time and can be used as a high-bandwidth signal akin to EEG [92]. However, EEG and fNIRs systems have different physical interfaces, sizes, weights and power budgets, thus different wearability and usability in real operative contexts. In this regard, the presence of hair may impact negatively on both photon absorption [93] and the coupling of the probes with the underlying scalp, thus the fNIRs technique is very reliable only on those un-hairy brain areas, like the *Pre Frontal Cortex* (PFC). For the mental states investigation, also other cortical regions, such as the parietal brain sites play an important role. Derosière et al. [94] pointed out how some fNIRs-measured hemodynamic variables were relatively insensitive to certain changes during the brain activity. In conclusion, due to its higher temporal resolution and usability, in comparison with the fNIRs technique, the EEG technique overcomes such issues related to the fNIRs and appears to be the better candidate for such kind of applications in operational environments. With particular regard to the mental workload literature, neurophysiological measurements have been and are often used to evaluate the level of cognitive demand induced by a task [16, 95, 96]. Most part of the EEG research showed that the brain electrical activities mainly considered for the mental workload analysis are the Theta and Alpha rhythms typically gathered from the *Pre-Frontal Cortex* (PFC) and the *Posterior Parietal Cortex* (PPC) regions. In this regard, previous studies demonstrated as that EEG Theta rhythm over the PFC present a positive correlation with mental workload [97, 98]. Moreover, published literature stressed the inverse correlation between the EEG power in the alpha frequency band over the PPC and mental workload [68, 99–103]. Only few studies have reported significant results about the modulation of the EEG power in other frequency bands, i.e. the delta, beta and gamma. Therefore, the most accepted evidences about EEG correlates of mental workload could be resumed in an increase of the theta band spectral power, especially on the frontal cortex, and a decrease in alpha band over the parietal cortex, with increasing mental workload [17, 68] (Fig. 2).

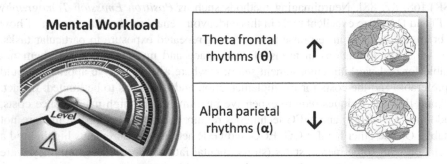

Fig. 2. Schematic summary of the main EEG features variations when the mental workload increasing.

Several studies, in particular in the aviation domain, have highlighted the high reliability of EEG-based mental workload indexes [99]. The results showed that the effects of task demand were evident on the EEG rhythms variations. EEG power spectra increased in the theta band, while significantly decreased in the alpha band as the task difficulty increased, over central, parietal, frontal and temporal brain sites. More recently, Shou and colleagues [104] evaluated mental workload during an ATC experiment using a new *time-frequency Independent Component Analysis* (tfICA) method for the analysis of the EEG signal. They found that "the frontal theta EEG activity was a sensitive and reliable metric to assess workload and time-on-task effect during an ATC task at the resolution of minute(s)". In other recent studies involving professional and trainees ATCOs [52, 105], it was demonstrated how it was possible to compute an EEG-based Workload Index able to significantly discriminate the workload demands of the ATM task by using machine-learning techniques and frontal-parietal brain features. In those studies, the ATM tasks were developed with a continuously varying difficulty levels in order to ensure realistic ATC conditions, i.e. starting form an easy level, then increasing up to a hard one and finishing with an easy one again. The EEG-based mental workload indices showed to be directly and significantly correlated with the actual mental demand experienced by the ATCOs during the entire task. However, the algorithms proposed were affected by some weaknesses, such as parameters manual settings and performance decreasing over time, that limited their employment in real operational environments. Moreover, other studies about mental workload estimation by using neurophysiological measurements, have been performed in other types of transport domain, in particular considering road transport (e.g. car drivers) [59, 68, 106, 107], and in the military domain [108].

3   Passive Brain-Computer Interfaces and Automation

Neuroergonomics research field aims at developing systems that take such limitations of a human's mental capacity to process information into account and avoid perfor-mance degradation, by adapting the user's interface to reduce the task demand/complexity or by intervening directly on the system [109]. Over the past two decades, researchers in the field of augmented cognition worked to develop novel technologies that can both monitor and enhance human cognition and performance. Most of this augmented cognition research was based on research findings coming from cognitive science and cognitive neuroscience. On the basis of such findings and the technological improvements, that have allowed to measure human biosignals in a more reliable and no-invasive way, it has been possible to evaluate the actual operator mental states by using neurophysiological indexes, and to use them as input toward the interface the operator is interacting. Such kind of application is called *passive Brain-Computer Interface* (passive-BCI). In its classical assumption, a Brain-Computer Interface (BCI) is a communication system in which messages or commands that an individual sends to the external world do not pass through the brain's normal output pathways of peripheral nerves and muscles [110]. More recently, Wolpaw and Wolpaw [111] defined a Brain-Computer Interface as "a system that measures Central Nervous System (CNS) activity and converts it into artificial output that replaces, restores, enhances, supplements, or improves natural CNS output and thereby changes the ongoing interactions between the CNS and its external or internal environment". In the BCI community, the possibility of using the BCI systems in different contexts for communication and system control [112, 113], developing also applications in eco-logical and operational environments, is not just a theory but something very closed to real applications [114–116]. In fact, in the classic BCI applications the user can modulate voluntarily its brain activity to interact with the system. In the new BCI concept, i.e. the passive BCI, the system recognizes the spontaneous brain activity of the user related to the considered mental state (e.g. emotional state, workload, attention levels), and uses such information to improve and modulate the interaction between the operator and the system itself [117]. Thus, in the context of Adaptive Automation (AA) in operational environments, the passive-BCI perfectly match the needs of the system in terms of *Human-Machine Interaction* (HMI) (Fig. 3).

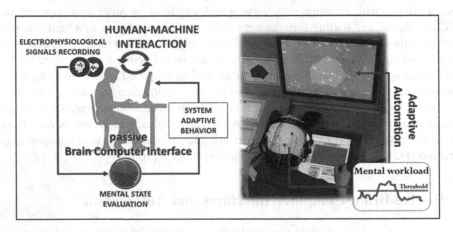

Fig. 3. Representation of the passive-BCI concept applied to enhance the Human-Machine Interaction by adapting the automation of an Air Traffic Management workstation. Source: https://doi.org/10.1109/TBME.2017.2694856 [117].

One of the main limitations of the use of EEG is its wearability. However technology improvements [75, 118–121] have being developed and tested in terms of dry electrodes (no gel and impedances adaptation issues), comfort, ergonomic and wireless communications (no cables between EEG sensors and the recording system). EEG-based passive-BCI systems appear the best candidate to be integrated in the development of AA-based systems and dynamically trigger the tasks allocation, on the basis of the user's actual mental state, i.e. his Mental Workload, in order to support him/her during his/her work activities consequently improving his/her performance, thus the safety of the whole environment. An issue that is still very much open is the development of a systematic method, in other words an algorithm, able to assess the user's Mental Workload online, despite all the problems related to operational environments (i.e. no controlled settings, artefacts, time cost in terms of calibration and computation, invasiveness on the subject, etc.), and in a way that is transferrable to various diverse environmental conditions.

4 Conclusions

Researchers in human factors and ergonomics sectors studied human capabilities and limitations, both cognitive and physical, and used such knowledge to design technologies and work environments to be safer and more usable, efficient, and enjoyable for people to interact with [7, 63, 122–124]. In today's technology-driven environment, where human capabilities are struggling to keep up with technology offerings, techniques for augmenting human performance are becoming the critical gap to preclude realizing the full benefits that these technology advances offer [125–134]. The concept of human performance augmentation is not so recent. The idea was developed during the past decade [64, 65], and, at the same time, the concept of Augmented Cognition (AugCog) was borne out of the Defense Advanced Research Projects Agency's (DARPA) pushing for technologies that enhanced the Warfighter's communication skills and those technologies that involved biosensors for medical applications [66]. More in general, because of the technological progresses, Human Factor research also evolved, and it now includes also the forefront techniques provided by Neuroscientific disciplines, that are looked upon with increasing interest by the scientific community and society in general. Alongside the latest technological development also the aims of workload assessment methods have evolved: the ultimate goal in particular is now geared towards Workload Adaptation, the process of workload management to aid learning, healing or limiting human errors. Moreover, workload measurement affects both the design and management of interfaces. On the one hand, by testing the workload of subjects during the use of web interfaces [122], for example, it is possible to direct the design. On the other hand, in the field of adaptive automation, it is the continuous monitoring of the workload level of the subject that allows the system to vary the feedback in response to the mental state of the operator [50, 71].

References

1. Wickens, C.D.: Processing resources in attention. In: Varieties of Attention, pp. 62–102 (1984)
2. Eggemeier, F.T., Wilson, G.F., Kramer, A.F., Damos, D.L.: Workload assessment in multi-task environments. In: Damos, D.L. (ed.) Multiple-Task Performance, pp. 207–216. Taylor & Francis, London (1991)
3. Hart, S.G., Staveland, L.E.: Development of NASA-TLX (Task Load Index): results of empirical and theoretical research. In: Hancock, P.A., Meshkati, N. (eds.) Advances in Psychology, vol. 52, pp. 139–183. North-Holland, Amsterdam (1988)
4. Gopher, D., Donchin, E.: Workload: an examination of the concept. In: Boff, K.R., Kaufman, L., Thomas, J.P. (eds.) Handbook of Perception and Human Performance. Cognitive Processes and Performance, vol. 2, pp. 1–49. John Wiley & Sons, Oxford (1986)
5. Cain, B.: A Review of the Mental Workload Literature, July 2007
6. Baldwin, C.L.: Commentary. Theor. Issues Ergon. Sci. 4(1–2), 132–141 (2003)
7. Wickens, C.D., Hollands, J.G., Banbury, S., Parasuraman, R.: Engineering Psychology & Human Performance, 4th edn. Psychology Press, Boston (2012)
8. Feyer, A., Williamson, A.M.: Human factors in accident modelling. In: Encyclopaedia of Occupational Health and Safety (2011)
9. Helmreich, R.L.: On error management: lessons from aviation. BMJ 320(7237), 781–785 (2000)
10. Brennan, T.A., et al.: Incidence of adverse events and negligence in hospitalized patients. Results of the Harvard Medical Practice Study I. N. Engl. J. Med. 324(6), 370–376 (1991)

11. Leape, L.L., et al.: The nature of adverse events in hospitalized patients. Results of the Harvard Medical Practice Study II. N. Engl. J. Med. **324**(6), 377–384 (1991)
12. Wilson, R.M., Runciman, W.B., Gibberd, R.W., Harrison, B.T., Newby, L., Hamilton, J. D.: The quality in Australian health care study. Med. J. Aust. **163**(9), 458–471 (1995)
13. Vincent, C., Neale, G., Woloshynowych, M.: Adverse events in British hospitals: preliminary retrospective record review. BMJ **322**(7285), 517–519 (2001)
14. Kohn, L.T., Corrigan, J.M., Donaldson, M.S.: To Err Is Human: Building a Safer Health System, vol. 6. National Academies Press, Washington, DC (2000)
15. Aberg, L., Rimmö, P.A.: Dimensions of aberrant driver behaviour. Ergonomics **41**(1), 39–56 (1998)
16. Arico, P., et al.: Human factors and neurophysiological metrics in air traffic control: a critical review. IEEE Rev. Biomed. Eng. **PP**(99), 1 (2017)
17. Borghini, G., Aricò, P., Di Flumeri, G., Babiloni, F.: Industrial Neuroscience in Aviation: Evaluation of Mental States in Aviation Personnel. BIOSYSROB, vol. 18. Springer, Cham (2017). https://doi.org/10.1007/978-3-319-58598-7
18. Salmon, P., Regan, M., Johnston, I.: Human Error and Road Transport: Phase One - Literature Review. Monash University Accident Research Centre (2005)
19. Shappel, S.A., Wiegmann, D.A.: The human factors analysis and classification system - HFACS. In: Federal Aviation Administration, Washington, DC, DOT/FAA/AM-00/7 (2000)
20. Stanton, N.A., et al.: Predicting design induced pilot error using HET (human error template) - A new formal human error identification method for flight decks, 1 February 2006. https://dspace.lib.cranfield.ac.uk/handle/1826/1158. Accessed 25 Nov 2015
21. Rankin, W., Hibit, R., Allen, J., Sargent, R.: Development and evaluation of the Maintenance Error Decision Aid (MEDA) process. Int. J. Ind. Ergon. **26**(2), 261–276 (2000)
22. Shorrock, S.T., Kirwan, B.: Development and application of a human error identification tool for air traffic control. Appl. Ergon. **33**(4), 319–336 (2002)
23. Edwards, T., Martin, L., Bienert, N., Mercer, J.: The relationship between workload and performance in air traffic control: exploring the influence of levels of automation and variation in task demand. In: Human Mental Workload: Models and Applications, pp. 120–139 (2017)
24. Lawton, R., Ward, N.J.: A systems analysis of the Ladbroke Grove rail crash. Accid. Anal. Prev. **37**(2), 235–244 (2005)
25. Reason, J.: Human error. West. J. Med. **172**(6), 393–396 (2000)
26. Rumar, K.: The basic driver error: late detection. Ergonomics **33**(10–11), 1281–1290 (1990)
27. Kirwan, B.: Human error identification techniques for risk assessment of high risk systems–Part 1: review and evaluation of techniques. Appl. Ergon. **29**(3), 157–177 (1998)
28. Sexton, J.B., Thomas, E.J., Helmreich, R.L.: Error, stress, and teamwork in medicine and aviation: cross sectional surveys. BMJ **320**(7237), 745–749 (2000)
29. Nelson, W.R., Haney, L.N., Ostrom, L.T., Richards, R.E.: Structured methods for identifying and correcting potential human errors in space operations. Acta Astronaut. **43** (3–6), 211–222 (1998)
30. Rasmussen, J.: Human errors. A taxonomy for describing human malfunction in industrial installations. J. Occup. Accid. **4**(2), 311–333 (1982)
31. Rasmussen, J.: The definition of human error and a taxonomy for technical system design. In: New Technology and Human Error, pp. 23–30 (1987)
32. Reason, J.: Human error: models and management. BMJ **320**(7237), 768–770 (2000)

33. Yerkes, R.M., Dodson, J.D.: The relation of strength of stimulus to rapidity of habit-formation. J. Comp. Neurol. Psychol. **18**(5), 459–482 (1908)
34. Westman, M., Eden, D.: The inverted-U relationship between stress and performance: a field study. Work Stress **10**(2), 165–173 (1996)
35. Rapolienė, L., Razbadauskas, A., Jurgelėnas, A.: The reduction of distress using therapeutic geothermal water procedures in a randomized controlled clinical trial. Adv. Prev. Med. (2015). https://www.hindawi.com/journals/apm/2015/749417/abs/. Accessed 1 Aug 2019
36. Calabrese, E.J.: Neuroscience and hormesis: overview and general findings. Crit. Rev. Toxicol. **38**(4), 249–252 (2008)
37. Warm, J.S., Parasuraman, R., Matthews, G.: Vigilance requires hard mental work and is stressful. Hum. Factors **50**(3), 433–441 (2008)
38. Fan, J., Smith, A.P.: The impact of workload and fatigue on performance. In: Longo, L., Leva, M.C. (eds.) H-WORKLOAD 2017. CCIS, vol. 726, pp. 90–105. Springer, Cham (2017). https://doi.org/10.1007/978-3-319-61061-0_6
39. Kirsh, D.: A few thoughts on cognitive overload. Intellectica **30**, 19–51 (2000)
40. Wickens, C.D.: Multiple resources and mental workload. Hum. Factors **50**(3), 449–455 (2008)
41. Wickens, C.D.: Mental workload: assessment, prediction and consequences. In: Longo, L., Leva, M.C. (eds.) H-WORKLOAD 2017. CCIS, vol. 726, pp. 18–29. Springer, Cham (2017). https://doi.org/10.1007/978-3-319-61061-0_2
42. Kantowitz, B.H., Casper, P.A.: Human workload in aviation. In: Human Error in Aviation, pp. 123–153. Routledge, London (2017)
43. Gerjets, P., Walter, C., Rosenstiel, W., Bogdan, M., Zander, T.O.: Cognitive state monitoring and the design of adaptive instruction in digital environments: lessons learned from cognitive workload assessment using a passive brain-computer interface approach. Front. Neurosci. **8**, 1–21 (2014)
44. Byrne, A.: Measurement of mental workload in clinical medicine: a review study. Anesthesiol. Pain Med. **1**(2), 90 (2011)
45. O'Donnell, R.D., Eggemeier, F.T.: Workload assessment methodology. In: Handbook of Perception and Human Performance, vol. 2. Wiley, New York (1986)
46. Orru, G., Longo, L.: The evolution of cognitive load theory and the measurement of its intrinsic, extraneous and Germane loads: a review. In: Longo, L., Leva, M.C. (eds.) H-WORKLOAD 2018. CCIS, vol. 1012, pp. 23–48. Springer, Cham (2019). https://doi.org/10.1007/978-3-030-14273-5_3
47. Scerbo, M.W.: Theoretical perspectives on adaptive automation. In: Parasuraman, R., Mouloua, M. (eds.) Automation and Human Performance: Theory and Applications, pp. 37–63. Lawrence Erlbaum Associates, Inc., Hillsdale (1996)
48. Parasuraman, R.: Neuroergonomics: research and practice. Theor. Issues Ergon. Sci. **4**(1–2), 5–20 (2003)
49. Tattersall, A.J., Foord, P.S.: An experimental evaluation of instantaneous self-assessment as a measure of workload. Ergonomics **39**(5), 740–748 (1996)
50. Aricò, P., Borghini, G., Di Flumeri, G., Colosimo, A., Pozzi, S., Babiloni, F.: A passive Brain-Computer Interface (p-BCI) application for the mental workload assessment on professional Air Traffic Controllers (ATCOs) during realistic ATC tasks. Prog. Brain Res. **228**, 295–328 (2016)
51. Borghini, G., et al.: Quantitative assessment of the training improvement in a motor-cognitive task by using EEG, ECG and EOG signals. Brain Topogr. **29**(1), 149–161 (2016)

52. Di Flumeri, G., et al.: On the use of cognitive neurometric indexes in aeronautic and air traffic management environments. In: Blankertz, B., Jacucci, G., Gamberini, L., Spagnolli, A., Freeman, J. (eds.) Symbiotic 2015. LNCS, vol. 9359, pp. 45–56. Springer, Cham (2015). https://doi.org/10.1007/978-3-319-24917-9_5
53. Mühl, C., Jeunet, C., Lotte, F.: EEG-based workload estimation across affective contexts. Neuroprosthetics **8**, 114 (2014)
54. Wierwille, W.W., Eggemeier, F.T.: Recommendations for mental workload measurement in a test and evaluation environment. Hum. Factors J. Hum. Factors Ergon. Soc. **35**(2), 263–281 (1993)
55. Endsley, M.R.: Measurement of situation awareness in dynamic systems. Hum. Factors **37**(1), 65–84 (1995)
56. Borghini, G., et al.: EEG-based cognitive control behaviour assessment: an ecological study with professional air traffic controllers. Sci. Rep. **7**(1), 547 (2017)
57. Dehais, F., et al.: Monitoring pilot's cognitive fatigue with engagement features in simulated and actual flight conditions using an hybrid fNIRS-EEG passive BCI. In: 2018 IEEE International Conference on Systems, Man, and Cybernetics (SMC), pp. 544–549 (2018)
58. Borghini, G., et al.: A new perspective for the training assessment: machine learning-based neurometric for augmented user's evaluation. Front. Neurosci. **11**, 325 (2017)
59. Di Flumeri, G., et al.: EEG-based mental workload neurometric to evaluate the impact of different traffic and road conditions in real driving settings. Front. Hum. Neurosci. **12**, 509 (2018)
60. Berka, C., et al.: Real-time analysis of EEG indexes of alertness, cognition, and memory acquired with a wireless EEG headset. Int. J. Hum. Comput. Interact. **17**(2), 151–170 (2004)
61. Dehais, F., Roy, R.N., Gateau, T., Scannella, S.: Auditory alarm misperception in the Cockpit: an eeg study of inattentional deafness. In: Schmorrow, D.D.D., Fidopiastis, C.M. M. (eds.) AC 2016, Part I. LNCS (LNAI), vol. 9743, pp. 177–187. Springer, Cham (2016). https://doi.org/10.1007/978-3-319-39955-3_17
62. McMahan, T., Parberry, I., Parsons, T.D.: Evaluating player task engagement and arousal using electroencephalography. Procedia Manuf. **3**, 2303–2310 (2015)
63. Cartocci, G., Maglione, A.G., Rossi, D., Modica, E., Borghini, G., Malerba, P., Piccioni, L. O., Babiloni, F.: Alpha and Theta EEG variations as indices of listening effort to be implemented in neurofeedback among cochlear implant users. In: Ham, J., Spagnolli, A., Blankertz, B., Gamberini, L., Jacucci, G. (eds.) Symbiotic 2017. LNCS, vol. 10727, pp. 30–41. Springer, Cham (2018). https://doi.org/10.1007/978-3-319-91593-7_4
64. Ahlstrom, U., Ohneiser, O., Caddigan, E.: Portable weather applications for general aviation pilots. Hum. Factors **58**(6), 864–885 (2016)
65. Giraudet, L., Imbert, J.-P., Bérenger, M., Tremblay, S., Causse, M.: The neuroergonomic evaluation of human machine interface design in air traffic control using behavioral and EEG/ERP measures. Behav. Brain Res. **294**, 246–253 (2015)
66. Prinzel, L.J., Freeman, F.G., Scerbo, M.W., Mikulka, P.J., Pope, A.T.: A closed-loop system for examining psychophysiological measures for adaptive task allocation. Int. J. Aviat. Psychol. **10**(4), 393–410 (2000)
67. Jänig, W.: Autonomic nervous system. In: Schmidt, R.F., Thews, G. (eds.) Human Physiology, pp. 333–370. Springer, Heidelberg (1989)
68. Borghini, G., Astolfi, L., Vecchiato, G., Mattia, D., Babiloni, F.: Measuring neurophysiological signals in aircraft pilots and car drivers for the assessment of mental workload, fatigue and drowsiness. Neurosci. Biobehav. Rev. **44**, 58–75 (2014)

69. VaezMousavi, S.M., Barry, R.J., Rushby, J.A., Clarke, A.R.: Arousal and activation effects on physiological and behavioral responding during a continuous performance task. Acta Neurobiol. Exp. (Warsz.) **67**(4), 461–470 (2007)
70. Sloan, R.P., et al.: Effect of mental stress throughout the day on cardiac autonomic control. Biol. Psychol. **37**(2), 89–99 (1994)
71. Aricò, P., et al.: Adaptive automation triggered by EEG-Based mental workload index: a passive brain-computer interface application in realistic air traffic control environment. Front. Hum. Neurosci. **10**, 539 (2016)
72. Cartocci, G., et al.: Gender and age related effects while watching TV advertisements: an EEG study. Comput. Intell. Neurosci. **2016**, 10 (2016)
73. Di Flumeri, G., et al.: EEG-based approach-withdrawal index for the pleasantness evaluation during taste experience in realistic settings. In: 2017 39th Annual International Conference of the IEEE Engineering in Medicine and Biology Society (EMBC), pp. 3228–3231 (2017)
74. Mihajlović, V., Grundlehner, B., Vullers, R., Penders, J.: Wearable, wireless EEG solutions in daily life applications: what are we missing? IEEE J. Biomed. Health Inform. **19**(1), 6–21 (2015)
75. Di Flumeri, G., Aricò, P., Borghini, G., Sciaraffa, N., Di Florio, A., Babiloni, F.: The dry revolution: evaluation of three different EEG dry electrode types in terms of signal spectral features, mental states classification and usability. Sensors **19**(6), 1365 (2019)
76. Izzetoglu, K., et al.: UAV operators workload assessment by optical brain imaging technology (fNIR). In: Valavanis, K.P., Vachtsevanos, G.J. (eds.) Handbook of Unmanned Aerial Vehicles, pp. 2475–2500. Springer, Dordrecht (2015). https://doi.org/10.1007/978-90-481-9707-1_22
77. Gateau, T., Ayaz, H., Dehais, F.: In silico versus over the clouds: On-the-fly mental state estimation of aircraft pilots, using a functional near infrared spectroscopy based passive-BCI. Front. Hum. Neurosci. **12**, 187 (2018)
78. Arico, P., et al.: Human factors and neurophysiological metrics in air traffic control: a critical review. IEEE Rev. Biomed. Eng. **10**, 250–263 (2017)
79. Borghini, G., et al.: Neurophysiological measures for users' training objective assessment during simulated robot-assisted laparoscopic surgery. In: 2016 IEEE 38th Annual International Conference of the Engineering in Medicine and Biology Society (EMBC), pp. 981–984 (2016)
80. Matthews, G., Reinerman-Jones, L.E., Barber, D.J., Abich IV, J.: The psychometrics of mental workload: multiple measures are sensitive but divergent. Hum. Factors **57**(1), 125–143 (2015)
81. Dehais, F., Peysakhovich, V., Scannella, S., Fongue, J., Gateau, T.: Automation surprise in aviation: real-time solutions. In: Proceedings of the 33rd Annual ACM Conference on Human Factors in Computing Systems, pp. 2525–2534 (2015)
82. Fallahi, M., Motamedzade, M., Heidarimoghadam, R.H., Soltanian, A.R., Miyake, S.: Effects of mental workload on physiological and subjective responses during traffic density monitoring: a field study. Appl. Ergon. **52**, 95–103 (2016)
83. Fallahi, M., Motamedzade, M., Heidarimoghadam, R., Soltanian, A.R., Miyake, S.: Assessment of operators' mental workload using physiological and subjective measures in cement, city traffic and power plant control centers. Health Promot. Perspect. **6**(2), 96 (2016)
84. Cherubino, P., et al.: neuroelectrical indexes for the study of the efficacy of TV advertising stimuli. In: Nermend, K., Łatuszyńska, M. (eds.) Selected Issues in Experimental Economics. SPBE, pp. 355–371. Springer, Cham (2016). https://doi.org/10.1007/978-3-319-28419-4_22

85. Di Flumeri, G., et al.: EEG frontal asymmetry related to pleasantness of olfactory stimuli in young subjects. In: Nermend, K., Łatuszyńska, M. (eds.) Selected Issues in Experimental Economics. SPBE, pp. 373–381. Springer, Cham (2016). https://doi.org/10.1007/978-3-319-28419-4_23

86. Kong, W., Lin, W., Babiloni, F., Hu, S., Borghini, G.: Investigating driver fatigue versus alertness using the granger causality network. Sensors 15(8), 19181–19198 (2015)

87. Ramnani, N., Owen, A.M.: Anterior prefrontal cortex: insights into function from anatomy and neuroimaging. Nat. Rev. Neurosci. 5(3), 184–194 (2004)

88. Wood, J.N., Grafman, J.: Human prefrontal cortex: processing and representational perspectives. Nat. Rev. Neurosci. 4(2), 139–147 (2003)

89. Cabeza, R., Nyberg, L.: Imaging cognition II: an empirical review of 275 PET and fMRI studies. J. Cogn. Neurosci. 12(1), 1–47 (2000)

90. Borghini, G., et al.: Frontal EEG theta changes assess the training improvements of novices in flight simulation tasks. Conf. Proc. IEEE Eng. Med. Biol. Soc. 2013, 6619–6622 (2013)

91. Ryu, K., Myung, R.: Evaluation of mental workload with a combined measure based on physiological indices during a dual task of tracking and mental arithmetic. Int. J. Ind. Ergon. 35(11), 991–1009 (2005)

92. Medvedev, A.V., Kainerstorfer, J., Borisov, S.V., Barbour, R.L., VanMeter, J.: Event-related fast optical signal in a rapid object recognition task: improving detection by the independent component analysis. Brain Res. 1236, 145–158 (2008)

93. Murkin, J.M., Arango, M.: Near-infrared spectroscopy as an index of brain and tissue oxygenation. Br. J. Anaesth. 103(Suppl. 1), i3–i13 (2009)

94. Derosière, G., Mandrick, K., Dray, G., Ward, T.E., Perrey, S.: NIRS-measured prefrontal cortex activity in neuroergonomics: strengths and weaknesses. Front. Hum. Neurosci. 7, 583 (2013)

95. Boucsein, W., Backs, R.W.: Engineering Psychophysiology: Issues and Applications. CRC Press, Boca Raton (2000)

96. Desmond, P.A., Hancock, P.A.: Active and Passive Fatigue States. Stress Workload Fatigue (2001)

97. Gevins, A., Smith, M.E.: Neurophysiological measures of cognitive workload during human-computer interaction. Theor. Issues Ergon. Sci. 4(1–2), 113–131 (2003)

98. Smit, A.S., Eling, P.A.T.M., Coenen, A.M.L.: Mental effort affects vigilance enduringly: after-effects in EEG and behavior. Int. J. Psychophysiol. Off. J. Int. Organ. Psychophysiol. 53(3), 239–243 (2004)

99. Brookings, J.B., Wilson, G.F., Swain, C.R.: Psychophysiological responses to changes in workload during simulated air traffic control. Biol. Psychol. 42(3), 361–377 (1996)

100. Gevins, A., Smith, M.E., McEvoy, L., Yu, D.: High-resolution EEG mapping of cortical activation related to working memory: effects of task difficulty, type of processing, and practice. Cereb. Cortex 7(4), 374–385 (1997)

101. Jaušovec, N., Jaušovec, K.: Working memory training: improving intelligence–changing brain activity. Brain Cogn. 79(2), 96–106 (2012)

102. Klimesch, W., Doppelmayr, M., Pachinger, T., Ripper, B.: Brain oscillations and human memory: EEG correlates in the upper alpha and theta band. Neurosci. Lett. 238(1–2), 9–12 (1997)

103. Venables, L., Fairclough, S.H.: The influence of performance feedback on goal-setting and mental effort regulation. Motiv. Emot. 33(1), 63–74 (2009)

104. Shou, G., Ding, L., Dasari, D.: Probing neural activations from continuous EEG in a real-world task: time-frequency independent component analysis. J. Neurosci. Methods 209(1), 22–34 (2012)

105. Arico, P., et al.: Towards a multimodal bioelectrical framework for the online mental workload evaluation. In: 2014 36th Annual International Conference of the IEEE Engineering in Medicine and Biology Society (EMBC), pp. 3001–3004 (2014)
106. Göhring, D., Latotzky, D., Wang, M., Rojas, R.: Semi-autonomous car control using brain computer interfaces. In: Lee, S., Cho, H., Yoon, K.J., Lee, J. (eds.) Intelligent Autonomous Systems 12. AISC, vol. 194, pp. 393–408. Springer, Heidelberg (2013). https://doi.org/10.1007/978-3-642-33932-5_37
107. Kohlmorgen, J., et al.: Improving human performance in a real operating environment through real-time mental workload detection (2007)
108. Dorneich, M.C., Ververs, P.M., Mathan, S., Whitlow, S.D.: A joint human-automation cognitive system to support rapid decision-making in hostile environments. In: 2005 IEEE International Conference on Systems, Man and Cybernetics, vol. 3, pp. 2390–2395 (2005)
109. Fuchs, S., Hale, K.S., Stanney, K.M., Juhnke, J., Schmorrow, D.D.: Enhancing mitigation in augmented cognition. J. Cogn. Eng. Decis. Mak. 1(3), 309–326 (2007)
110. Wolpaw, J.R., Birbaumer, N., McFarland, D.J., Pfurtscheller, G., Vaughan, T.M.: Brain-computer interfaces for communication and control. Clin. Neurophysiol. Off. J. Int. Fed. Clin. Neurophysiol. 113(6), 767–791 (2002)
111. Wolpaw, J.W., Wolpaw, E.W.: Brain-Computer Interfaces: Principles and Practice. Oxford University Press, Oxford (2012)
112. Aloise, F., et al.: A covert attention P300-based brain-computer interface: Geospell. Ergonomics 55(5), 538–551 (2012)
113. Riccio, A., et al.: Hybrid P300-based brain-computer interface to improve usability for people with severe motor disability: electromyographic signals for error correction during a spelling task. Arch. Phys. Med. Rehabil. 96(3 Suppl.), S54–S61 (2015)
114. Blankertz, B., et al.: The Berlin brain–computer interface: non-medical uses of BCI technology. Front. Neurosci. 4, 198 (2010)
115. Müller, K.-R., Tangermann, M., Dornhege, G., Krauledat, M., Curio, G., Blankertz, B.: Machine learning for real-time single-trial EEG-analysis: from brain-computer interfacing to mental state monitoring. J. Neurosci. Methods 167(1), 82–90 (2008)
116. Zander, T.O., Kothe, C., Welke, S., Roetting, M.: Utilizing secondary input from passive brain-computer interfaces for enhancing human-machine interaction. In: Schmorrow, D.D., Estabrooke, I.V., Grootjen, M. (eds.) FAC 2009. LNCS (LNAI), vol. 5638, pp. 759–771. Springer, Heidelberg (2009). https://doi.org/10.1007/978-3-642-02812-0_86
117. Aricò, P., Borghini, G., Di Flumeri, G., Sciaraffa, N., Colosimo, A., Babiloni, F.: Passive BCI in operational environments: insights, recent advances, and future trends. IEEE Trans. Biomed. Eng. 64(7), 1431–1436 (2017)
118. Chi, Y.M., Wang, Y.-T., Wang, Y., Maier, C., Jung, T.-P., Cauwenberghs, G.: Dry and noncontact EEG sensors for mobile brain-computer interfaces. IEEE Trans. Neural Syst. Rehabil. Eng. 20(2), 228–235 (2012)
119. Liao, L.-D., et al.: Biosensor technologies for augmented brain-computer interfaces in the next decades. Proc. IEEE 100, 1553–1566 (2012). Special Centennial Issue
120. Lopez-Gordo, M.A., Sanchez-Morillo, D., Valle, F.P.: Dry EEG electrodes. Sensors 14(7), 12847–12870 (2014)
121. Borghini, G., Aricò, P., Di Flumeri, G., Sciaraffa, N., Babiloni, F.: Correlation and Similarity between cerebral and non-cerebral electrical activity for user's states assessment. Sensors 19(3), 704 (2019)
122. Jimenez-Molina, A., Retamal, C., Lira, H.: Using psychophysiological sensors to assess mental workload during web browsing. Sens. Switz. 18(2), 1–26 (2018)
123. Longo, L., Leva, M.C. (eds.): H-WORKLOAD 2017. CCIS, vol. 726. Springer, Cham (2017). https://doi.org/10.1007/978-3-319-61061-0

124. Longo, L., Leva, M.C. (eds.): H-WORKLOAD 2018. CCIS, vol. 1012. Springer, Cham (2019). https://doi.org/10.1007/978-3-030-14273-5
125. Leva, C., Wilkins, M., Coster, F.: Human performance modelling for adaptive automation (2018)
126. Dearing, D., Novstrup, A., Goan, T.: Assessing workload in human-machine teams from psychophysiological data with sparse ground truth. In: Longo, L., Leva, M.C. (eds.) H-WORKLOAD 2018. CCIS, vol. 1012, pp. 13–22. Springer, Cham (2019). https://doi.org/10.1007/978-3-030-14273-5_2
127. Junior, A.C., Debruyne, C., Longo, L., O'Sullivan, D.: On the mental workload assessment of uplift mapping representations in linked data. In: Longo, L., Leva, M.C. (eds.) H-WORKLOAD 2018. CCIS, vol. 1012, pp. 160–179. Springer, Cham (2019). https://doi.org/10.1007/978-3-030-14273-5_10
128. Comberti, L., Leva, M.C., Demichela, M., Desideri, S., Baldissone, G., Modaffari, F.: An empirical approach to workload and human capability assessment in a manufacturing plant. In: Longo, L., Leva, M.C. (eds.) H-WORKLOAD 2018. CCIS, vol. 1012, pp. 180–201. Springer, Cham (2019). https://doi.org/10.1007/978-3-030-14273-5_11
129. Longo, L.: Experienced mental workload, perception of usability, their interaction and impact on task performance. PLoS ONE **13**(8), e0199661 (2018)
130. Byrne, A.: Mental workload as an outcome in medical education. In: Longo, L., Leva, M.C. (eds.) H-WORKLOAD 2017. CCIS, vol. 726, pp. 187–197. Springer, Cham (2017). https://doi.org/10.1007/978-3-319-61061-0_12
131. Longo, L.: Designing medical interactive systems via assessment of human mental workload. In: 2015 IEEE 28th International Symposium on Computer-Based Medical Systems, pp. 364–365. IEEE, June 2015
132. Rizzo, L., Longo, L.: Representing and inferring mental workload via defeasible reasoning: a comparison with the NASA task load index and the workload profile. In: 1st Workshop on Advances in Argumentation in Artificial Intelligence, AI^3@AI*IA, Bari, Italy, pp. 126–140 (2017)
133. Moustafa, K., Longo, L.: Analysing the impact of machine learning to model subjective mental workload: a case study in third-level education. In: Longo, L., Leva, M.C. (eds.) H-WORKLOAD 2018. CCIS, vol. 1012, pp. 92–111. Springer, Cham (2019). https://doi.org/10.1007/978-3-030-14273-5_6
134. Rizzo, L., Longo, L.: Inferential models of mental Workload with defeasible argumentation and non-monotonic fuzzy reasoning: a comparative study. In: 2nd Workshop on Advances in Argumentation in Artificial Intelligence, Trento, Italy, pp. 11–26 (2019)

Real-Time Mental Workload Estimation Using EEG

Aneta Kartali[1,2], Milica M. Janković[2(✉)], Ivan Gligorijević[1(✉)],
Pavle Mijović[1(✉)], Bogdan Mijović[1(✉)], and Maria Chiara Leva[3(✉)]

[1] mBrainTrain LLC, Požarevačka 36, 11000 Belgrade, Serbia
{aneta.kartali,ivan,pavle.mijovic,
bogdan}@mbraintrain.com
[2] School of Electrical Engineering, University of Belgrade,
Bulevar kralja Aleksandra 73, 11000 Belgrade, Serbia
piperski@etf.rs
[3] School of Environmental Health, Technological University Dublin,
Kevin Street, Dublin 2, Ireland
mariachiara.leva@tudublin.ie

Abstract. Tracking mental workload in real-time during worker's performance is a challenge, as it requires the worker to report during the task execution. Moreover, it is thus based on subjective experience. Neuroergonomics tackles this issue, by employing neurophysiological metrics to obtain objective, real-time information. We measured mental workload (MWL) derived from electroencephalography (EEG) signals, for subjects engaged in a simulated computer-based airplane-landing task. To test this metric, we calculated the degree of correlation between measured MWL and observable variables associated to task complexity. In the two settings of the experiment, we used a 24-channel full-cap EEG system and the novel mobile EEG headphone device. The latter allows seamless integration of the EEG acquisition system in a possible real-world setup scenario. Obtained results reveal significant correlation between the EEG derived MWL metric and the two objective task complexity metrics: the number of airplanes on the screen subjects had to control, as well as the number of actions performed by the subject during the task in both setups. Therefore, this work represents a proof of concept for using the proposed systems for reliable real-time mental workload tracking.

Keywords: EEG · Mental workload · Neuroergonomics

1 Introduction

Mental workload (MWL) is defined as a measure of human ability to retain focus and rational reasoning while processing multiple activities and facing distracting influences [1]. MWL is generally considered to be correlated with task demand and performance. Studies showed that excessive, as well as low mental workload, can degrade task performance and cause errors [2, 3]. Therefore, there is an increasing need to quantify mental workload in real-time, in order to determine its optimal level and hence improve one's efficiency at work. Unfortunately, one of the major difficulties encountered in

© Springer Nature Switzerland AG 2019
L. Longo and M. C. Leva (Eds.): H-WORKLOAD 2019, CCIS 1107, pp. 20–34, 2019.
https://doi.org/10.1007/978-3-030-32423-0_2

studying mental load is its measurement. Conventionally, mental workload is evaluated through subjective and objective scores [4]. Subjective measures are obtained from individuals' subjective estimations of task difficulty and their overall perceived experience of the task [5]. On the other hand, objective measures generally include metrics such as task score and accuracy. However, both subjective and objective workload measures have certain drawbacks, one of which is their inability to provide real-time and continuous information. Subjective measures are derived from subject's self-report so this kind of data is gathered after the task is finished or the task must be paused for the report to be made. Also, people are often biased when they report on their own experiences [6]. When it comes to objective measures such as accuracy or task score, they can too be obtained only after the task is finished. Furthermore, they might not even indicate the experienced load, since this depends on the nature of the task. Recent studies have been focused on validating the use of physiological responses for quantifying the MWL of individuals [7–9]. It has been shown that neurophysiological measurements such as electroencephalography (EEG) signals are directly correlated with mental demand experienced during the task [10]. Namely, certain EEG spectral components vary in a predictable way in response to the cognitive demands of the task [11–13]. This means that correlation exists between EEG spectral power and task complexity: the increase in frontal midline theta band (4–7 Hz) and a decrease in parietal midline alpha band (8–12 Hz), have been observed when the task complexity increases. A ratio between these two power bands is proven to be a reliable index for MWL estimation [14]. It is shown that EEG measures of MWL correlate both with subjective and objective performance metrics [15]. The need for a high temporal resolution, unobtrusive acquisition and obtaining reliable and accurate MWL measures make EEG advantageous compared to other neurophysiological measurement techniques [16]. Despite their benefits, traditional EEG recording systems imply wired data acquisition, which causes significant degradation in signal quality, mainly due to the movement artifacts. Another drawback of the traditional EEG systems are their size and weight, which constrains them to laboratory environments [17]. With the recent development of wearable EEG systems, mobility and compactness that came with wireless signal transmission made EEG recording possible in various, every-day settings [18, 19]. These characteristics suggest that an EEG based MWL estimator can be applicable in ergonomic studies (research field known as the Neuroergonomics [20]), where workers' brain responses can be recorded in the workplace [19–23]. Another step forward in such environments would be the use of concealed EEG equipment to reduce real-world interruptions that are treated as psychological noise. Large part of such noise stems from the fact that random passersby are "attracted" by the EEG cap, which consequently make the subjects feel uncomfortable, distracting their attention from the experimental task. This can further raise an issue of validity of a desired experiment. With the concept of concealed EEG, integrated in headphones, we mitigate this problem and demonstrate the ability to conduct outside-of-lab cognitive experiments. In this paper we addressed a real-time measurement of MWL in the relevant, out-of-the-lab setting, while participants played an airplane landing game. EEG data were acquired using a mobile 24-channel full-cap EEG system, as well as a newly developed EEG equipment – Smartfones, where the EEG device is integrated in the headphones, containing a restricted set of 11 electrodes. Such a device allows seamless integration of the acquisition system in a possible real-world setup scenario. We correlated EEG derived MWL metrics with task complexity measures and hypothesized positive correlation between these modalities. Finally, we compared the results obtained using the traditional full-cap EEG system and the EEG headphones and found that the EEG system integrated in the headphones can be used for measuring real-time mental workload in complex, real-world scenarios.

2 Workload Assessment Method and Experimental Design

2.1 Experimental Protocol

Eighteen healthy volunteers (8 female and 10 male) of age 26.89 ± 4.29 participated in the study and were all asked to sign a written consent. Intake of alcohol, caffeine and demanding physical activity were restricted prior to the beginning of the experiments. The experimental environment consisted of a quiet office room with adjusted light intensity and room temperature. The conditions were maintained during different days of experimentation. There were two recording sessions which were sometimes run on different days. One session was recorded with the full-cap 24-channel EEG system (Smarting, mBrainTrain LLC, Belgrade, Serbia) and the other with the EEG head-phones (Smartfones, mBrainTrain LLC, Belgrade Serbia), while half of randomly selected subjects were recorded with the cap first and then with the headphones and the other half were recorded in the reversed order. Each session consisted of three trials, where the first trial was the training trial and the second trial was used as the experiment task. For some participants that finished the game too soon during the second trial (with three or less airplanes on the screen), the additional, third trial was used. Prior to every session, two-minute baseline signal was recorded during which the subject was instructed to comfortably sit and relax.

2.2 Experimental Task

Subjects were instructed to play the "airplane landing" game developed by the U.K. National Air Traffic Services (NATS) [24], while their EEG was recorded. A scene example of the game is shown in Fig. 1. Player's task consisted of directing the airplanes, represented by yellow circles, to the black runway. Airplanes can be navigated by clicking on the white arrows located on the rings concentric to the yellow circles. The game is over when two yellow circles collide, and the player is warned when their outer rings overlap. Number of airplanes gradually increases and so does their speed. That way, task complexity increases as the subject must navigate a greater number of airplanes at the same time. Experimental run was recorded from the beginning of the game, until the airplanes collide. Experimental data were recorded using Smarting Streamer for PC (mBrainTrain LLC, Belgrade, Serbia) which is compatible with the Lab Streaming Layer (LSL) [25]. LSL collects and synchronizes multiple time series inputs across devices via the LAN network. The experiment setup is shown in Fig. 2. Two PCs connected via Ethernet cable were used in the setup. The study participants played the game on one PC, while the other one was used for the data acquisition. During the experiments, three streams were recorded: EEG stream sent via

Bluetooth connection to the recording PC and two trigger streams, which were all written together into an XDF format file [26]. One trigger stream gave information about the number of airplanes currently visible on the screen. To provide this information, the experimenter manually inserted a trigger marker from the keyboard, every time the number of airplanes on the screen changed. The second trigger stream was the mouse click trigger stream (we named this stream *interaction event*) sent from the playing computer via Ethernet cable, using the LSL protocol. Every time when a subject clicks on the screen during the game, a trigger is received by the Streamer, visualized and written in an XDF file, together with the EEG stream and the airplane-number trigger stream.

Fig. 1. ATC landing game scene during the experiment. On the top in the center is the runway where the airplanes should be directed. Airplanes are represented as yellow circles with concentric rings with white arrows used for their navigation. In the presented screen, there are five airplanes. (Color figure online)

Fig. 2. Experiment setup. *Left:* Experimenter inserting airplane-number trigger stream on the data acquisition computer. *Right:* Subject playing the "airplane landing" game [24] and wearing the headphones for EEG measurement.

2.3   EEG Measurements

In one recording session a 24-channel gel-based EEG cap with 10–20 electrode placement (Fig. 3(d)) and the Ag/AgCl electrodes was used (EASYCAP GmbH, Wörthsee, Germany, Fig. 3(b)) [27]. For the other session we used Smartfones (mBrainTrain LLC, Belgrade, Serbia), headphones with the electrodes to record EEG signals [28]. This configuration uses saline based (gel-free) electrodes (Greentek, Wuhan, China) placed around the ears (electrode placement is depicted in Fig. 3(c)), with the additional three electrodes located in the central scalp zone (C_3, C_4 and Cz electrodes according to the 10–20 system, Fig. 3(a)). In both configurations, EEG data were acquired using 500 Hz sampling frequency. When the gel-based system was used, the electrode impedances were kept below 10 kΩ, whereas in the second configuration they were kept below 20 kΩ because of different electrode properties.

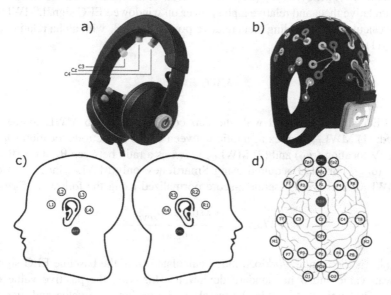

Fig. 3. (*a*) Smartfones; (*b*) EEG cap with the Smarting amplifier; (*c*) Earpad electrode positions on Smartfones; (*d*) Electrode positions on the cap.

2.4   EEG Data Preprocessing

EEG data were processed using MATLAB (MathWorks, Massachusetts, USA) and EEGLAB [29]. Channels highly contaminated with noise were excluded prior to the processing procedure. The EEG signals were first band-pass filtered in the range 1–40 Hz, using a FIR filter generated by EEGLAB. Further processing included *Artifact Subspace Reconstruction* (ASR) algorithm, a recently developed method for eliminating high amplitude noise, based on the approach of signal reconstruction using a reference signal fragment [30]. It can automatically eliminate undesired artifacts such as eye blinks, muscle and cable movement artifacts which makes the processing procedure convenient for real-time use. Re-referencing to electrodes placed on the mastoid bones was applied afterwards. Full-cap EEG data was re-referenced to the average of mastoid electrodes (M_1, M_2), and signals recorded using Smartfones were re-referenced to the average of L_4, R_4 electrodes (see Fig. 3).

2.5 EEG Workload Metrics

The following analysis included the estimation of relative band powers of the windowed data and calculation of the MWL metrics. The EEG signal was decomposed into frequency components by applying Fast Fourier transform (FFT). First, the EEG signal was windowed using a rectangular window (3 s length, with overlap of 2.9 s) and an estimate of the power spectral density (PSD) was obtained for every windowed segment This procedure was followed by calculation of the relative band power from the signal segment theta and alpha frequency bands, which were determined using *Individual Alpha Frequency* (IAF) [31]. The overall results are two time series which represent relative theta and relative alpha power on windowed EEG signal. MWL index was then calculated by dividing theta relative power signal θ_x with alpha relative power signal α_x (1).

$$MWL = \frac{\theta_x}{\alpha_x} \tag{1}$$

For EEG signals acquired with the cap configuration, two MWL indices were calculated: (1) MWL_{FzPz} index as a ratio between θ_x at Fz electrode location (θ_{Fz}) and α_x at the Pz location (α_{Pz}) and (2) MWL_{Cz} index as a ratio between θ_x at Cz (θ_{Cz}) and α_x at Cz (α_{Cz}). For EEG acquired using Smartfones, only MWL_{Cz} index was calculated. MWL indices were baseline z-score normalized using the following formula

$$MWL_{z-score} = \frac{MWL - \mu_{MWL_{baseline}}}{\sigma_{MWL_{baseline}}} \tag{2}$$

$MWL_{baseline}$ is the mental workload index calculated from the baseline EEG signal, μ is its mean value and σ its standard deviation. This way, the positive value of the $MWL_{z-score}$ indicates that workload rises above its resting state value and vice versa, while its amplitude represents the workload intensity, relative to the resting state value.

2.6 Task Complexity Metrics

To validate the use of physiological measure of MWL it should be compared to certain objective task complexity measures. In that respect, we used two different task complexity metrics. The first complexity metric represents the changing number of airplanes on the screen and was obtained from a keyboard trigger stream (see Sect. 2.2) – we call this metric *the number of airplanes*. It is a step function which magnitude corresponds to the current number of airplanes on the screen, at any given time, during the game. The second metric of task complexity is the number of subject's interactions with the game in a given time interval, obtained from the *interaction events* stream (see Sect. 2.2). To compute the correlation between the MWL metric and the interaction events, these events were averaged using a moving average filter (3 s length, overlap 2.9 s) to obtain the same number of samples between the different signal modalities.

2.7 Statistical Analysis

Statistical analysis was applied to validate the used physiological measure of mental workload. In that respect, we calculated Pearson's correlation between $MWL_{z\text{-score}}$ and each of the two task complexity metrics. To calculate the correlation between $MWL_{z\text{-score}}$ and the step signal representing the changing number of airplanes, an average of $MWL_{z\text{-score}}$ is obtained from every interval in which the airplane number is constant. For calculating correlation with *the interaction events*, $MWL_{z\text{-score}}$ signal was averaged on intervals with the constant mean *interaction events* values. We calculated and compared correlations using $MWL_{z\text{-score}}$ derived from the θ_{Fz}/α_{Pz} ratio and $MWL_{z\text{-score}}$ derived solely from the Cz electrode site (θ_{Cz}/α_{Cz} ratio). Furthermore, we compared these metrics with $MWL_{z\text{-score}}$ calculated from the Smartfones EEG data (as a θ_{Cz}/α_{Cz} ratio).

3 Interim Results

A graphical representation of the processed data is shown in Fig. 4 (when the full-cap system was used for data acquisition) and Fig. 5 (when the Smartfones were used for data acquisition). These example data belong to the subject number 4 from the study. MWL metric was calculated as a ratio between central theta power band and central alpha power band, MWL_{Cz}. For the purposes of visualization, $MW_{z\text{-score}}$ signal was smoothed using a 6th-order Butterworth lowpass filter with the cutoff frequency of $\pi/8$ rad/sample and further multiplied by a scalar which doesn't influence the correlation results.

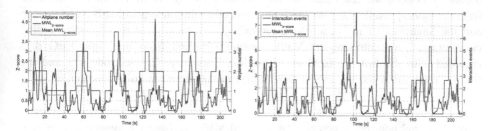

Fig. 4. Full-cap configuration, Cz-based MWL metric. *Left*: *Number of airplanes*, smoothed $MWL_{z\text{-score}}$ signal, and averaged $MWL_{z\text{-score}}$ on the intervals where the airplane number is constant; MA windowed $MWL_{z\text{-score}}$ and; *Right*: MA windowed *interaction events*, smoothed $MWL_{z\text{-score}}$ signal, and averaged $MWL_{z\text{-score}}$ on the intervals with the constant mean *interaction events* values. (Color figure online)

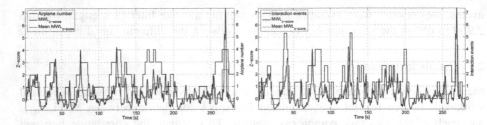

Fig. 5. Smartfones configuration, Cz-based MWL metric: *Left: Number of airplanes*, smoothed MWL$_{z\text{-score}}$ signal, and averaged MWL$_{z\text{-score}}$ on the intervals where the airplane number is constant; MA windowed MWL$_{z\text{-score}}$ and; *Right*: MA windowed *interaction events*, smoothed MWL$_{z\text{-score}}$ signal, and averaged MWL$_{z\text{-score}}$ on the intervals with the constant mean *interaction events* values. (Color figure online)

Correlation results between the proposed MWL metric and each of the two task complexity metrics, are shown in Tables 1 and 2. For the cap recordings, four out of eighteen subjects had their second trial replaced with the third, due to the early game over. These subjects were subject number 1, 4, 8 and 13. For the Smartfones recordings, this was done for five subjects, number 1, 9, 13, 15 and 18. Table 1 shows Pearson's correlations between the averaged MWL metric and *the number of airplanes*, using the full-cap configuration and the Smartfones. Table 2 shows Pearson's correlations between the averaged MWL metric MA windowed *interaction events*. For the full-cap configuration results are represented using the MWL metric calculated as the θ_{Fz}/α_{Pz} ratio (MWL$_{FzPz}$) and as the θ_{Cz}/α_{Cz} ratio (MWL$_{Cz}$), while the results obtained using Smartfones MWL metric was calculated as the θ_{Cz}/α_{Cz} ratio (MWL$_{Cz}$). Statically significant correlation is marked with a (*) if the corresponding p-value is lower than 0.05. Average values of statistically significant correlations are shown in Table 3. In order to check if solely the Cz electrode location can be used for the estimation of the mental workload, we calculated the correlation between the MWL metrics obtained from the θ_{Fz}/α_{Pz} and θ_{Cz}/α_{Cz}. As the Cz electrode has the same position both on the cap and the Smartfones and because one recording session implied using one recording configuration, we compare the MWL metrics obtained from the same EEG data recorded with the full-cap EEG configuration. This way, Cz electrode site and the θ_{Cz}/α_{Cz} ratio simulated the use of Smartfones in the same recording session.

Table 1. Correlation results between the averaged MWL$_{z\text{-score}}$ and *the number of airplanes*, using the full-cap EEG and Smartfones recording configuration. (* - p < 0.05)

	Full-cap EEG system		Smartfones
	θ_{Fz}/α_{Pz}	θ_{Cz}/α_{Cz}	θ_{Cz}/α_{Cz}
subject 1	0.39*	0.42*	0.20*
subject 2	−0.25*	−0.18	−0.39*
subject 3	0.39*	0.44*	0.43*

(continued)

Table 1. (*continued*)

| | Full-cap EEG system | | Smartfones |
	θ_{Fz}/α_{Pz}	θ_{Cz}/α_{Cz}	θ_{Cz}/α_{Cz}
subject 4	0.46*	0.44*	0.28*
subject 5	0.41*	0.45*	−0.23*
subject 6	0.44*	0.33*	0.51*
subject 7	0.36*	0.17	0.24*
subject 8	0.13	0.18	−0.37*
subject 9	0.18	0.15	0.01
subject 10	−0.23	−0.23	0.07
subject 11	0.23	0.16	−0.06
subject 12	0.29*	0.30*	0.19*
subject 13	0.30*	0.16*	0.14
subject 14	0.28*	0.33*	0.22
subject 15	0.40*	0.36*	0.32*
subject 16	0.58*	0.39*	0.24*
subject 17	0.34*	0.17	0.35*
subject 18	0.41*	0.24	−0.40*

Table 2. Correlation results between the averaged $MWL_{z-score}$ and MA windowed *interaction events*, using the full-cap EEG and Smartfones recording configuration. (* - $p < 0.05$)

| | Full-cap EEG system | | Smartfones |
	θ_{Fz}/α_{Pz}	θ_{Cz}/α_{Cz}	θ_{Cz}/α_{Cz}
subject 1	0.27*	0.33*	0.27*
subject 2	0.02	0.05	−0.29*
subject 3	−0.11	0.34*	0.45*
subject 4	0.42*	0.43*	0.30*
subject 5	0.24*	0.17*	−0.06
subject 6	0.33*	0.37*	0.45*
subject 7	0.34*	0.21*	0.29*
subject 8	0.22*	0.15*	−0.14
subject 9	0.15	0.06	0.09
subject 10	0.08	0.01	0.05
subject 11	0.26*	0.12	0.02
subject 12	0.22*	0.18*	0.08
subject 13	0.26*	0.22*	0.17*
subject 14	0.13	0.14	0.26*
subject 15	0.12	0.02	0.30*
subject 16	0.43*	0.21	0.12
subject 17	0.34*	0.32*	0.38*
subject 18	0.36*	0.12	−0.16

Table 3. Mean values and standard deviations of statistically significant correlation results.

Averaged MWL – *number of airplanes* correlation mean values and standard deviations		
Full-cap EEG system		Smartfones
θ_{Fz}/α_{Pz}	θ_{Cz}/α_{Cz}	θ_{Cz}/α_{Cz}
(0.30 ± 0.23) *	(0.35 ± 0.09) *	(0.11 ± 0.33) *
Averaged MWL – MA windowed *interaction events* correlation mean values and standard deviations		
Full-cap EEG system		Smartfones
θ_{Fz}/α_{Pz}	θ_{Cz}/α_{Cz}	θ_{Cz}/α_{Cz}
(0.31 ± 0.07) *	(0.27 ± 0.1) *	(0.26 ± 0.21) *

Figure 6 represents the correlation results between the EEG MWL metrics obtained from different scalp locations (Fz and Pz versus solely Cz). These correlations are all statistically significant (p < 0.05) for all subjects and are reasonably high.

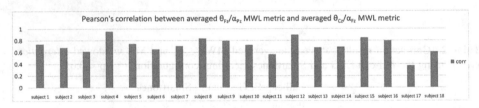

Fig. 6. Correlations between averaged MWL_{FzPz} and MWL_{Cz} obtained from the same EEG data recorded with the full-cap EEG configuration.

4 The Implications of Real Time Mental Workload Assessment for Safety Critical Domains

As socio-technical systems are becoming even more complex, i.e. subject to deeply uncertain conditions and dynamism, it is not possible to assure a perfect fit in human-machine interaction (HMI) just by acting at the design phase, when a finite set of possible future states of the system are postulated [32]. The system will change over time and the human element will have to change accordingly to optimally reconfigure the HMI and assure higher performance [33]. So the questions to be addressed are more and more concerned with the possibility to support the operator in this dynamic search for a new fit in HMI that accommodates human capabilities (e.g. workload, comfort, performance, etc.) as well as the system's mission under new context conditions not in a rigid design-operation sequence but real time, supporting the base for an adaptive automaton "learning" feature in HMI. The research in this field needs to bring together a multidisciplinary capacity where neuroergonomics can play a revolutionary role if it can operate in the context of highly automated, safety-critical domain, and will substantially improve the ability of industry to address the difficult problems faced in designing human interactions for highly automated systems. Mental workload is a variable closely connected with Human-System Performance. Previous works [34, 35] have successfully assessed/predicted human error (considered in terms of the correct or incorrect execution of a task). The approach was based on the use of a generalised linear regression with a Poisson mode with the assumption that Human Performance can be represented as directly dependent from two macro-factors:

- Task complexity (TC): summarising all factors contributing to physical and mental workload requirements for execution of a given operative task.
- Human capability (HC): summarising the skills, training and experience of the people facing the tasks, representing a synthesis of their physical and cognitive abilities to verify whether or not they match the task requirements.

The model is represented in Eq. (3):

$$\lambda_i = e^{\beta_0 + \beta_1 x_1 - \beta_2 x_2 + \varepsilon_i} = e^{\eta_i} \rightarrow log\lambda_i = \eta_i \qquad (3)$$

where λ_i is the amount of error recorded, x_1 is a metric representing observable metrics of task complexity features and x_2 is a variable representing operator skill level/capability to cope with the situation. The study reported in this paper explored the use of EEG as a tool to track individual human capabilities of the subjects in real-time (as individually perceived mental workload metric) and it was tested on a task related to an "airplane landing" serious game and compared to objective metrics associated to task complexity (number of airplanes on the screen and action events undertaken by the participant). To do so, the MWL metric proposed in the literature [14], which is based on calculating the ratio between frontal midline theta power band and parietal midline alpha power band (MWL = θ_{Fz}/α_{Pz}) was correlated with two objective task complexity metrics (*the number of airplanes and route conflicts/interaction events*). The results show statically significant correlations between the proposed MWL EEG derived metric and both task complexity metrics for fourteen out of the eighteen tested subjects (Figs. 1 and 2). These findings are a promising step towards validating the use of the proposed EEG MWL metric to assess worker's mental load during the task execution in real time. *The interaction events* metric tells us more about how much the subject is immersed in the task at a specific moment, compared to *the number of airplanes*, which is simply an objective metric observed by the experimenter. However, this metric provided weaker correlations than *the number of airplanes* metric, indicating it may not be a good indicator of the mental workload. Yet, other objective metrics, similar to this one, which are potentially a better indicator of mental workload should still be explored in the future. In addition, we explored the potential of estimating MWL using a single electrode site (such as Cz which is located on the central scalp region). This verification is an important step towards the possibility of using this index in real-life application, where the number of electrode sites is restricted due to equipment limitations. For that reason, we tested whether it is possible to obtain similar results using this single electrode site for MWL estimation (Cz). Hence, we first processed the EEG signals recorded with the full-cap system, calculated the MWL metric as a θ_{Cz}/α_{Cz} ratio and compared it with the θ_{Fz}/α_{Pz} MWL metric. The obtained correlation results were high (r = 0.71 ± 0.15) and highly statistically significant (p < 0.01) for all the participants in the study (see Fig. 6). Moreover, both EEG based MWL metrics (θ_{Fz}/α_{Pz} and θ_{Cz}/α_{Cz}) provide similar results when correlated with task complexity metrics, as shown in Figs. 1 and 2. Therefore, it can be concluded that MWL metrics calculated solely from the central electrode site (Cz) represents a viable alternative for the MWL estimation. Correlation results of the θ_{Cz}/α_{Cz} MWL metric obtained using Smartfones are significant for a smaller number of subjects, compared to the full-cap setting. This can be due to one or both of the following reasons. First, the session when the MWL is measured with Smartfones was not the same as the session when MWL is measured with the cap. For some participants, sessions were recorded on different days. Second, the full-cap system uses gel-based electrodes, contrary to semi-dry electrodes, used with Smartfones. However, gel-based electrodes are not well-suited for real-life applications, and therefore, we consider that the results obtained using Smartfones are more relevant in this respect. The third reason lies in the fact that the Smartfones we used for recording were still in the prototyping stage, and therefore not all the electrodes provided good contacts for all the participants throughout the experiment recordings. Therefore, this study is only the proof of concept, and further studies and repeated experiments are necessary in order to obtain reliable results and conclusions. In most of the studies found in the literature [36–39], results obtained from estimating MWL are evaluated and compared on the group level [10, 11, 15, 16] and MWL estimation is usually based on a classification or a regression model, in which case the result highly depends on the diversity of the training data. In this study we based the MWL estimation on an individual level, which meant that individual differences were considered rather than obtaining only the group results. This can set the path for the development of an unobtrusive EEG-based MWL estimator which will be sensitive to inter-individual varieties. This study opens the road to the possibility of comparing the subjective capability of coping with different scenarios that can be calibrated to estimate thresholds where errors are likely to occur. When the human error probability prediction real time scores above a certain threshold, the model can be used to suggest measures to support decision making for the operator and/or call for external supervisory help (and providing the supervisor with an easy to grasp quick summary of status). This is an example of an adaptive automation feature that can be tested and generated on the bases of the possibility opened by this combined real-time workload and performance assessment approach.

5 Conclusion

This study investigated the validity of physiological MWL indicator derived from EEG. Our aim was to test the reliability of an MWL estimator, during the task whose complexity gradually increases. We hypothesized that the increase of task complexity would induce more mental load which can be seen in the EEG and which we confirmed by calculating the correlation between the EEG derived MWL metrics and two objective task complexity metrics. Furthermore, we investigated the possibility to use the Smartfones for the MWL estimation, as they have a restricted number of channels and are missing the Fz and Pz electrodes, which are mostly used for the MWL estimation from the EEG signal. Investigating the possibility to use the θ_{Cz}/α_{Cz} ratio as an alternative for the reliable measurement of MWL, we provide the results which confirm that Smartfones can be used for this purpose. Although further studies are needed to estimate the precision and reliability of such a system with more precision, this is the first study to confirm the concept that Smartfones can be used as a tool for unobtrusive, real-time MWL tracking in the workplace. As the proposed EEG system can be used for real-time unobtrusive MWL estimation, further research could investigate its advancements with the use of machine learning algorithms and their abilities to reliably predict workers' overload without interfering with their work.

References

1. Recarte, M.A., Nunes, L.M.: Mental workload while driving: effects on visual search, discrimination, and decision making. J. Exp. Psychol. Appl. **9**(2), 119 (2003)
2. Gawron, V.J.: Human Performance Measures Handbook. Lawrence Erlbaum Associates Publishers, New Jersey (2000)
3. Paas, F.G., Van Merriënboer, J.J.: Variability of worked examples and transfer of geometrical problem-solving skills: a cognitive-load approach. J. Educ. Psychol. **86**(1), 122 (1994)
4. Yeh, Y.Y., Wickens, C.D.: Dissociation of performance and subjective measures of workload. Hum. Factors **30**(1), 111–120 (1988)
5. Reid, G.B., Nygren, T.E.: The subjective workload assessment technique: a scaling procedure for measuring mental workload. Adv. Psychol. **52**, 185–218 (1988)
6. Peterson, D.A., Kozhokar, D.: Peak-end effects for subjective mental workload ratings. Proc. Hum. Factors Ergon. Soc. Ann. Meet. **61**(1), 2052–2056 (2017)
7. Kapoor, A., Burleson, W., Picard, R.W.: Automatic prediction of frustration. Int. J. Hum. Comput. Stud. **65**(8), 724–736 (2007)
8. Gevins, A., Smith, M.E.: Neurophysiological measures of cognitive workload during human-computer interaction. Theor. Issues Ergon. Sci. **4**(1–2), 113–131 (2003)
9. Kramer, A.F.: Physiological metrics of mental workload: a review of recent progress. In: Damos, D.L. (ed.) Multiple-Task Performance, pp. 279–328. Taylor and Francis, London (1991)
10. Brookings, J.B., Wilson, G.F., Swain, C.R.: Psychophysiological responses to changes in workload during simulated air traffic control. Biol. Psychol. **42**(3), 361–377 (1996)
11. Gevins, A., Smith, M.E., McEvoy, L., Yu, D.: High-resolution EEG mapping of cortical activation related to working memory: effects of task difficulty, type of processing, and practice. Cereb. Cortex **7**(4), 374–385 (1997)
12. Missonnier, P., et al.: Frontal theta event-related synchronization: comparison of directed attention and working memory load effects. J. Neural Transm. **113**(10), 1477–1486 (2006)
13. Stipacek, A., Grabner, R.H., Neuper, C., Fink, A., Neubauer, A.C.: Sensitivity of human EEG alpha band desynchronization to different working memory components and increasing levels of memory load. Neurosci. Lett. **353**(3), 193–196 (2003)
14. Holm, A., Lukander, K., Korpela, J., Sallinen, M., Müller, K.M.: Estimating brain load from the EEG. Sci. World J. **9**, 639–651 (2009)
15. Berka, C., et al.: EEG correlates of task engagement and mental workload in vigilance, learning, and memory tasks. Aviat. Space Environ. Med. **78**(5), B231–B244 (2007)

16. Hogervorst, M.A., Brouwer, A.M., Van Erp, J.B.: Combining and comparing EEG, peripheral physiology and eye-related measures for the assessment of mental workload. Front. Neurosci. **8**, 322 (2014)

17. Debener, S., Minow, F., Emkes, R., Gandras, K., De Vos, M.: How about taking a low-cost, small, and wireless EEG for a walk? Psychophysiology **49**(11), 1617–1621 (2012)

18. De Vos, M., Kroesen, M., Emkes, R., Debener, S.: P300 speller BCI with a mobile EEG system: comparison to a traditional amplifier. J. Neural Eng. **11**(3), 036008 (2014)

19. Mijović, P.,Ković, V., De Vos, M., Mačužić, I., Todorović, P., Jeremić, B.: Towards continuous and real-time attention monitoring at work: reaction times versus brain response. Ergonomics **60**(2), 241–254 (2017)

20. Parasuraman, R., Rizzo, M.: Neuroergonomics: The Brain at Work. Oxford University Press, New York (2008)

21. Mijović, P., Milovanović, M.,Ković, V., Gligorijević, I., Mijović, B., Mačužić, I.: Neuroergonomics method for measuring the influence of mental workload modulation on cognitive state of manual assembly worker. In: Longo, L., Leva, M.C. (eds.) H-WORKLOAD 2017. CCIS, vol. 726, pp. 213–224. Springer, Cham (2017). https://doi.org/10.1007/978-3-319-61061-0_14

22. Kramer, A.F., Parasuraman, R.: Neuroergonomics: applications of neuroscience to human factors. In: Cacciopo, J., Tassinary, L.G., Berntson, G.G. (eds.) Handbook of Psychophysiology. Cambridge University Press, New York (2005)

23. Mijović, P.,Ković, V., De Vos, M., Mačužić, I., Jeremić, B., Gligorijević, I.: Benefits of instructed responding in manual assembly tasks: an ERP approach. Front. Hum. Neurosci. **10**, 171 (2016)

24. NATS games: ATC landing. https://www.nats.aero/careers/trainee-air-traffic-controllers/games/game/atclanding/. Accessed Aug 2019

25. Lab Streaming Layer. https://github.com/sccn/labstreaminglayer. Accessed Aug 2019

26. XDF: TheExtensible Data Format Based on XML Concepts. https://nssdc.gsfc.nasa.gov/nssdc_news/june01/xdf.html. Accessed Aug 2019

27. https://www.mbraintrain.com/wp-content/uploads/2016/01/RBE-24-STD.pdf.   Accessed Aug 2019

28. mBrainTrain: Smartfones. https://mbraintrain.com/smartfones/. Accessed Aug 2019

29. Swartz Center for Computational Neuroscience: EEGLAB. https://sccn.ucsd.edu/eeglab/index.php. Accessed Aug 2019

30. Mullen, T.R., et al.: Real-time neuroimaging and cognitive monitoring using wearable dry EEG. IEEE Trans. Biomed. Eng. **62**(11), 2553–2567 (2015)

31. Klimesch, W.: EEG alpha and theta oscillations reflect cognitive and memory performance: a review and analysis. Brain Res. Rev. **29**(2–3), 169–195 (1999)

32. Park, J., Seager, T.P., Rao, P.S.C., Convertino, M., Linkov, I.: Integrating risk and resilience approaches to catastrophe management in engineering systems. Risk Anal. **33**(3), 356–367 (2013)

33. Dekker, S.: The Field Guide to Understanding Human Error. Ashgate, Hampshire (2016)

34. Comberti, L., Leva, M.C., Demichela, M., Desideri, S., Baldissone, G., Modaffari, F.: An empirical approach to workload and human capability assessment in a manufacturing plant. In: Longo, L., Leva, M.C. (eds.) H-WORKLOAD 2018. CCIS, vol. 1012, pp. 180–201. Springer, Cham (2019). https://doi.org/10.1007/978-3-030-14273-5_11

35. Leva, M.C., Caimo, A., Duane, R., Demichela, M., Comberti, L.: Task complexity, and operators' capabilities as predictor of human error: modeling framework and an example of application. In: Haugen, S., Barros, A., van Gulijk, C., Kongsvik, T., Vinnem, J.E. (eds.) Safety and Reliability–Safe Societies in a Changing World: Proceedings of ESREL. CRC Press, Boca Raton (2018)

36. Longo, L.: Subjective usability, mental workload assessments and their impact on objective human performance. In: Bernhaupt, R., Dalvi, G., Joshi, A., Balkrishan, D.K., O'Neill, J., Winckler, M. (eds.) INTERACT 2017. LNCS, vol. 10514, pp. 202–223. Springer, Cham (2017). https://doi.org/10.1007/978-3-319-67684-5_13
37. Rizzo, L., Longo, L.: Representing and inferring mental workload via defeasible reasoning. In: Proceedings of the 1st Workshop on Advances in Argumentation in Artificial Intelligence (2017)
38. Longo, L., Leva, M.C. (eds.): H-WORKLOAD 2018. CCIS, vol. 1012. Springer, Cham (2019). https://doi.org/10.1007/978-3-030-14273-5
39. Longo, L., Leva, M.C. (eds.): H-WORKLOAD 2017. CCIS, vol. 726. Springer, Cham (2017). https://doi.org/10.1007/978-3-319-61061-0

Student Workload, Wellbeing and Academic Attainment

Andrew P. Smith[✉]

Centre for Occupational and Health Psychology, School of Psychology,
Cardiff University, 63 Park Place, Cardiff CF10 3AS, UK
smithap@cardiff.ac.uk

Abstract. There has been extensive research on workload, often in the laboratory or workplace. Less research has been conducted in educational settings and there is very little examining workload, wellbeing and academic attainment of university students. The present study of 1294 students examined associations between perceptions of workload, hours spent at university, time pressure and attainment and wellbeing outcomes (measured using the Wellbeing Process Questionnaire). Established predictors (stressors; social support; negative coping; positive personality and conscientiousness) were controlled for, and the analyses showed that workload was significantly associated with all outcomes whereas time pressure was only related to course stress and negative wellbeing (life stress, fatigue and anxiety/depression). Hours spent at the university had no significant effects. The effects of workload were interpreted in terms of an initial challenge leading to increased efficiency and attainment. These results show the importance of including workload in future longitudinal research on student wellbeing and attainment.

Keywords: Workload · Wellbeing · Academic attainment · Time pressure

1 Introduction

Mental workload has been investigated using a variety of different approaches [1, 2], and it has a long history in Psychology and Ergonomics [3, 4]. It has been examined in both laboratory [5, 6] and occupational settings [7, 8], and a variety of measures have been proposed [9–14]. These include self-assessment, task measures and physiological measures.

Self-assessment measures or subjective measures have taken several forms such as the NASA Task Load Index [15], the Workload Profile [16] and the Subjective Workload Assessment Technique [1]. Recent research has shown that even single items about perceptions of workload are highly correlated with longer scales and can predict wellbeing of workers [7, 8]. Other approaches have examined specific aspects of workload, such as time pressure. This is a major feature of the Karasek Job Demands scale, which has been shown to predict health and safety outcomes of workers [17].

Student workload research has the potential to lead educators and key stakeholders to best practices in teaching, reduce academic stress, and decrease college student dropout rates. Identifying best practices regarding student workload issues has the

© Springer Nature Switzerland AG 2019
L. Longo and M. C. Leva (Eds.): H-WORKLOAD 2019, CCIS 1107, pp. 35–47, 2019.
https://doi.org/10.1007/978-3-030-32423-0_3

potential for better outcomes in student learning. Despite the potential importance of studying student workload, there is little literature on students' workload, with a search of PsychInfo only revealing 16 articles. These were often concerned with the planning of the curriculum [18–22] or the relationship between assessment frequency and workload [23]. Other research has examined workload in distance learning [18, 24] and attempted to determine whether workload changes approaches to learning [20, 25] or how different teaching styles influence workload [26]. The present study is part of a programme of research examining factors which influence the wellbeing and academic attainment of students [27–34]. Again, the literature on these topics is very small with 3 articles on workload and attainment being identified [21, 26, 35], no articles on workload and wellbeing of students, and very few on related topics such workload and student performance [21] or workload and student stress [19, 22]. The aim of the present study was to provide information on workload, well-being and attainment using the Student Wellbeing Questionnaire [27].

Recent research has investigated well-being at work using occupational predictors (e.g. demands and resources) and individual effects (e.g. personality and coping). This research has been based on the Demands-Resources-Individual Effects model (DRIVE Model [36]), which allows for the inclusion of new predictors and outcomes. The model was initially designed to study occupational stress, and initial research [37, 38] focused on predictors of anxiety and depression. More recently, it has been adapted to study wellbeing, and recent studies [39–42] have also investigated predictors of positive well-being (happiness, positive affect and job satisfaction). When one studies both positive and negative aspects of wellbeing, then the number of predictors and outcomes grow, which can lead to a very long questionnaire. Our research has developed single or short item questions which have been shown to have the same predictive validity as multi-item scales. The original scale (the Well-being Process Questionnaire, WPQ) has been used with different occupational groups (e.g. nurses and university staff [41, 42]). This has led to the development of another questionnaire (the Smith Well-being Questionnaire – SWELL [7, 8]), which measures a wider range of predictors (e.g. the working environment and hours of work) and outcomes (e.g. absenteeism; presenteeism; sick leave; performance efficiency; work-life balance and illness caused or made worse by work).

The wellbeing of university students has been widely studied [43], and high levels of depression, anxiety, and stress have been reported by undergraduate students [44, 45]. These variables were, therefore, included in the Student WPQ. Positive and negative wellbeing are not just opposite ends of a single dimension, which meant that positive outcomes (happiness, life satisfaction and positive affect) were also included in the questionnaire. Student related stressors have been widely studied and include fear of failing and long hours of study [43], social demands [44–46] and lack of social support [47]. Questionnaires have been specifically developed to audit factors that influence well-being (e.g. the Inventory of College Students' Recent Life Experiences [ICSRLE] which measures time pressures, challenges to development, and social mistreatment [48]. The Student WPQ includes a short version of the ICSRLE and the most relevant question in this study was the one asking about time pressure. Research on students' well-being has also shown the importance of individual differences such as negative coping style and positive personality (high self-efficacy, high self-esteem and high optimism) in the well-being process. Smith [28] demonstrated that fatigue was an important part of the wellbeing process and research with occupational samples has shown strong links between fatigue and workload. Williams et al. [27] also demonstrated that the WPQ could predict students' reports of cognitive function and Smith [28] showed that this also applies to objective measures of academic attainment.

The original WPQ also included questions related to both life in general and to specifically to academic issues. One of the academic questions related to the perception of workload. This was intended to be an indicator of habitual workload rather than reflecting acute peaks and troughs. Others were concerned with hours spent at the university, course stress and efficiency of working. The aim of the present study was to examine associations between workload, time pressure, hours at the university, and the general positive and negative wellbeing outcomes. Academic attainment and perception of work efficiency were also obtained, and consideration of course stress allowed for analysis of specific academic challenges. One of the most important feature of the study was that established predictors of wellbeing (student stressors; negative coping; positive personality and social support) and attainment (conscientiousness) were statistically adjusted for.

2 Method

2.1 Participants

The participants were 1299 first and second year undergraduate Psychology students at Cardiff University (89.4% female; mean age = 19.4 years, range = 18–46, 98% 18–22 years). The study was approved by the Ethics Committee, School of Psychology, Cardiff University, and carried out with the informed consent of the participants. At the end of the survey the participants were shown a debrief statement and awarded course credits for their participation.

2.2 The Survey

The survey was presented online using the *Qualtrics* software.

2.3 Questions

The full set of questions are shown in Appendix A.1. The majority were taken from the version of the WPQ used with workers. The additional questions measured student stressors, aspects of perceived social support and university workload, time pressure, hours in university, course stress and work efficiency.

2.4 Derived Variables from the Survey

Five control variables were derived:

- Stressors (sum of ICSRLE questions)
- Social support (sum of ISEL questions)

- Positive personality (optimism + self-esteem + self-efficacy)
- Negative coping (avoidance; self-blame; wishful thinking)
- Conscientiousness (single question)

These variables were used because previous research showed that they were established predictors of the wellbeing outcomes. Other variables (e.g. the Big 5 personality scores) were not used here due to their lack of sensitivity in our previous studies.

Three measures of workload were used:

- Hours in university for taught courses
- Perception of workload (rated on a 10 point scale)
- Time pressure (rated on a 10 point scale).

Four outcome variables were derived:

- Positive outcomes (happiness + positive affect + life satisfaction)
- Negative outcomes (anxiety + depression + stress)
- Stress due to academic issues
- Efficiency of working

2.5 Academic Attainment

Students gave permission for their academic attainment scores to be made available and combined with their survey data (after which the database was anonymised). The score used here, the Grade Point Average, reflected the combination of coursework and examination marks.

3 Results

The mean score for perceived workload was 7.3 (s.d. = 3.2; higher scores = greater workload; possible range = 1 to 10). Similarly, the mean score for time pressure was 7.74 (s.d. = 1.73; higher scores = greater workload; possible range = 1 to 10). The mean number of hours spent in university was 9.8 h a week (s.d. = 3.2).

Initial univariate correlations showed that workload and time pressure were significantly correlated (r = 0.35 p < 0.001) but the hours in university variable was not correlated with the other two indicators of workload. The correlations between the workload measures and the wellbeing and attainment outcomes are shown in Table 1.

Table 1. Correlations between workload measures and wellbeing and attainment outcomes.

	Time pressure	Hours in university	Workload
GPA	−0.02	−0.06	0.04
Work efficiency	−0.61*	0.01	0.14*
Course stress	0.48**	0.08**	0.60**
Negative wellbeing	0.31**	0.02	0.19**
Positive wellbeing	−0.14**	0.01	0.00

*p < 0.05, **p < 0.01

Time pressure was significantly correlated with all of the outcomes except the GPA score. Hours in university was only correlated with course stress, and workload was correlated with all of the variables except GPA and positive wellbeing. However, these univariate analyses do not take into account the impact of the established predictors of the outcomes, or the shared variance with the other measures of workload. The workload scores were dichotomised into high/low groups and entered as independent variables into a MANOVA. The dependent variables were GPA, work efficiency, course stress, negative wellbeing and positive wellbeing. The co-variates were the other stressors, social support, negative coping, positive personality and conscientiousness.

The MANOVA revealed significant overall effects of workload (Wilks' Lambda = 0.792 F = 65.3 df 5, 1243 p < 0.001) and time pressure (Wilks' Lambda = 0.932 F = 18.2 df = 5, 1243 p < 0.001) but not hours in university. There were no significant interactions between the independent variables. Examination of the individual dependent variables showed that workload had significant effects on all of them (see Table 2; work efficiency: F 1,1247 = 13.6 p < 0.001; course stress: F 1, 1247 = 294.9 p < 0.001; GPA F 1,1247 = 2.6 p < 0.05 1-tail; negative wellbeing: F 1, 1247 = 10.4 p < 0.001; positive wellbeing: F 1, 1247 = 9.4 p < 0.005) whereas time pressure only had effects on course stress and negative wellbeing (life stress, anxiety and depression – see Table 3 – course stress: F 1,1247 = 82.8 p < 0.001; negative wellbeing: F 1,1247 = 14.67 p < 0.001). Those who reported a high workload had higher work efficiency scores, a higher GPA, more positive wellbeing but also greater course stress and negative wellbeing. Those with high time pressure scores reported greater course stress and higher negative wellbeing scores.

Table 2. Effects of workload on wellbeing and attainment outcomes (scores are the means, s.e.s in parentheses)

	Work efficiency	Course stress	GPA	Negative wellbeing	Positive wellbeing
Low workload	5.76 (0.70)	6.23 (0.053)	63.18 (0.28)	19.48 (0.19)	19.30 (0.10)
High workload	6.15 (0.08)	7.61 (0.06)	63.88 (0.32)	20.39 ((0.21)	19.74 (0.11)

Table 3. Effects of time pressure on course stress and negative wellbeing (scores are the means, s.e.s in parentheses)

	Course stress	Negative wellbeing
Low time pressure	6.54 (0.07)	19.37 (0.23)
High time pressure	7.30 (0.05)	20.50 (0.09)

4 Discussion

The main aim of the present study was to examine the associations between various single measures of workload and wellbeing and academic attainment of university students. The sample was very homogeneous in that it largely consisted of female Psychology students doing similar courses and of a similar age. A key feature of the present study was that established predictors of wellbeing (stressors; negative coping; positive personality and social support) and academic attainment (conscientiousness) were statistically controlled. Smith [49] has also shown that workload load and time pressure are associated with established predictors, which means that one has to control for these other predictors when examining the association between workload and wellbeing.

The analyses showed that hours spent at university had little influence on wellbeing and attainment, which probably reflects the fact that students do a great deal of their university work at home. Indeed, lectures are now filmed and can be watched online away from the university, which means that it is not necessary to attend them. A better question would have been to ask how many hours students spent on their academic studies.

Time pressure had selective effects, with high time pressure being linked with greater course stress and higher negative well-being scores. Such results are in agreement with research on workers where high demands, often induced by time pressure, are associated with higher perceived stress and reduced wellbeing [7, 8]. In contrast, time pressure had no significant effect on academic attainment (GPA scores) or perceive work efficiency.

Perceived workload had a more global effect on the outcome measures, although it appeared to have both positive and negative effects. For example, high workload was associated with greater work efficiency, higher GPA scores and higher positive well-being. In contrast, high workload was also associated with greater course stress and higher negative wellbeing. It is possible that the positive and negative effects of workload occur at different time points. It is plausible that increased stress may have a motivating effect which then improves efficiency and academic marks, which suggests that high workload is a challenge rather than a threat and this increases achievement motivation.

4.1   Limitations

The present study has two main limitations. The first is that it is a cross-sectional study and the results could reflect reverse causality with the outcomes changing perceptions of workload and time pressure rather than effects occurring in the opposite direction. Longitudinal studies, preferably with interventions changing workload, are required to determine whether workload has direct effects on wellbeing and attainment. The second problem is that while a homogenous sample means that one need not control for factors such as age or culture, these variables may represent important influences that should be considered. For example, Omosehin and Smith (in press) examined the effects of time pressure on the wellbeing of students differing in ethnicity and culture. The results for the group as a whole largely confirmed the effects reported here, but there were also significant interactions between ethnic group and time pressure, which demonstrated the importance of conducting cross-cultural research. Finally, the study was not designed to address underlying mechanisms, and further research is required to address the microstructure of workload and time pressure. Also, the applicability of classic inverted-U models of workload and the notion of optimal state needs to be examined.

5 Conclusion

In summary, the present study investigated whether single-item measures of workload were predictors of wellbeing and attainment outcomes. Established predictors were also measured, as these have been shown to be associated with both the independent and dependent variables. The results showed that hours at university had little effect whereas both time pressure and ratings of workload had significant associations. Time pressure was associated with course stress and negative wellbeing (life stress, depression and anxiety). Workload had a more global effect, increasing course stress and negative wellbeing, but also being associated with higher positive wellbeing, work efficiency and GPA scores. The effects of workload may reflect a challenge rather than a threat. The workload may initially be perceived as stressful, but the associated increased motivation may then lead to greater efficiency and attainment. Further research with a longitudinal design is needed to test this view, but at the moment, one can conclude that workload is a variable that should be added to the Student WPQ.

A Appendix

A.1 Students' Well-Being Questionnaire

The following questions contain a number of single-item measures of aspects of your life as a student and feelings about yourself. Many of these questions will contain examples of what thoughts/behaviours the question is referring to which are important for understanding the focus of the question, but should be regarded as guidance rather than strict criteria. Please try to be as accurate as possible, but avoid thinking too much about your answers, your first instinct is usually the best.

1.	Overall, I feel that I have low self-esteem (For example: At times, I feel that I am no good at all, at times I feel useless, I am inclined to feel that I am a failure) Disagree strongly 1 2 3 4 5 6 7 8 9 10 Agree strongly
2.	On a scale of one to ten, how depressed would you say you are in general? (e.g. feeling 'down', no longer looking forward to things or enjoying things that you used to) Not at all depressed 1 2 3 4 5 6 7 8 9 10 Extremely depressed
3.	I have been feeling good about my relationships with others (for example: Getting along well with friends/colleagues, feeling loved by those close to me) Disagree strongly 1 2 3 4 5 6 7 8 9 10 Agree strongly
4.	I don't really get on well with people (For example: I tend to get jealous of others, I tend to get touchy, I often get moody) Disagree strongly 1 2 3 4 5 6 7 8 9 10 Agree strongly
5.	Thinking about myself and how I normally feel, in general, I mostly experience positive feelings (For example: I feel alert, inspired, determined, attentive) Disagree strongly 1 2 3 4 5 6 7 8 9 10 Agree strongly

(continued)

(*continued*)

6.	In general, I feel optimistic about the future (For example: I usually expect the best, I expect more good things to happen to me than bad, It's easy for me to relax) Disagree strongly 1 2 3 4 5 6 7 8 9 10 Agree strongly
7.	I am confident in my ability to solve problems that I might face in life (For example: I can usually handle whatever comes my way, If I try hard enough I can overcome difficult problems, I can stick to my aims and accomplish my goals) Disagree strongly 1 2 3 4 5 6 7 8 9 10 Agree strongly
8.	I feel that I am laid-back about things (For example: I do just enough to get by, I tend to not complete what I've started, I find it difficult to get down to work) Disagree strongly 1 2 3 4 5 6 7 8 9 10 Agree strongly
9.	I am not interested in new ideas (For example: I tend to avoid philosophical discussions, I don't like to be creative, I don't try to come up with new perspectives on things) Disagree strongly 1 2 3 4 5 6 7 8 9 10 Agree strongly
10.	Overall, I feel that I have positive self-esteem (For example: On the whole I am satisfied with myself, I am able to do things as well as most other people, I feel that I am a person of worth) Disagree strongly 1 2 3 4 5 6 7 8 9 10 Agree strongly
11.	I feel that I have the social support I need (For example: There is someone who will listen to me when I need to talk, there is someone who will give me good advice, there is someone who shows me love and affection) Disagree strongly 1 2 3 4 5 6 7 8 9 10 Agree strongly
12.	Thinking about myself and how I normally feel, in general, I mostly experience negative feelings (For example: I feel upset, hostile, ashamed, nervous) Disagree strongly 1 2 3 4 5 6 7 8 9 10 Agree strongly
13.	I feel that I have a disagreeable nature (For example: I can be rude, harsh, unsympathetic) Disagree strongly 1 2 3 4 5 6 7 8 9 10 Agree strongly

Negative Coping Style:
Blame Self

14.	When I find myself in stressful situations, I blame myself (e.g. I criticize or lecture myself, I realise I brought the problem on myself). Disagree strongly 1 2 3 4 5 6 7 8 9 10 Agree strongly

Wishful Thinking

15.	When I find myself in stressful situations, I wish for things to improve (e.g. I hope a miracle will happen, I wish I could change things about myself or circumstances, I daydream about a better situation). Disagree strongly 1 2 3 4 5 6 7 8 9 10 Agree strongly

Avoidance

16.	When I find myself in stressful situations, I try to avoid the problem (e.g. I keep things to myself, I go on as if nothing has happened, I try to make myself feel better by eating/drinking/smoking). Disagree strongly 1 2 3 4 5 6 7 8 9 10 Agree strongly

(*continued*)

(continued)

Personality	
17.	I prefer to keep to myself (For example: I don't talk much to other people, I feel withdrawn, I prefer not to draw attention to myself) Disagree strongly 1 2 3 4 5 6 7 8 9 10 Agree strongly
18.	I feel that I have an agreeable nature (For example: I feel sympathy toward people in need, I like being kind to people, I'm co-operative) Disagree strongly 1 2 3 4 5 6 7 8 9 10 Agree strongly
19.	In general, I feel pessimistic about the future (For example: If something can go wrong for me it will, I hardly ever expect things to go my way, I rarely count on good things happening to me) Disagree strongly 1 2 3 4 5 6 7 8 9 10 Agree strongly
20.	I feel that I am a conscientious person (For example: I am always prepared, I make plans and stick to them, I pay attention to details) Disagree strongly 1 2 3 4 5 6 7 8 9 10 Agree strongly
21.	I feel that I can get on well with others (For example: I'm usually relaxed around others, I tend not to get jealous, I accept people as they are) Disagree strongly 1 2 3 4 5 6 7 8 9 10 Agree strongly
22.	I feel that I am open to new ideas (For example: I enjoy philosophical discussion, I like to be imaginative, I like to be creative) Disagree strongly 1 2 3 4 5 6 7 8 9 10 Agree strongly
23.	Overall, I feel that I am satisfied with my life (For example: In most ways my life is close to my ideal, so far I have gotten the important things I want in life) Disagree strongly 1 2 3 4 5 6 7 8 9 10 Agree strongly
24.	On a scale of one to ten, how happy would you say you are in general? Extremely unhappy 1 2 3 4 5 6 7 8 9 10 Extremely happy
25.	On a scale of one to ten, how anxious would you say you are in general? (e.g. feeling tense or 'wound up', unable to relax, feelings of worry or panic) Not at all anxious 1 2 3 4 5 6 7 8 9 10 Extremely anxious
26.	Overall, how stressful is your life? Not at all stressful 1 2 3 4 5 6 7 8 9 10 Very Stressful

Please consider the following elements of student life and indicate overall to what extent they have been a part of your life over the past 6 months. Remember to use the examples as guidance rather than trying to consider each of them specifically:

27.	Challenges to your development (e.g. important decisions about your education and future career, dissatisfaction with your written or mathematical ability, struggling to meet your own or others' academic standards). Not at all part of my life 1 2 3 4 5 6 7 8 9 10 Very much part of my life
28.	Time pressures (e.g. too many things to do at once, interruptions of your school work, a lot of responsibilities). Not at all part of my life 1 2 3 4 5 6 7 8 9 10 Very much part of my life
29.	Academic Dissatisfaction (e.g. disliking your studies, finding courses uninteresting, dissatisfaction with school). Not at all part of my life 1 2 3 4 5 6 7 8 9 10 Very much part of my life
30.	Romantic Problems (e.g. decisions about intimate relationships, conflicts with boyfriends'/girlfriends' family, conflicts with boyfriend/girlfriend). Not at all part of my life 1 2 3 4 5 6 7 8 9 10 Very much part of my life

(continued)

(continued)

31.	Societal Annoyances (e.g. getting ripped off or cheated in the purchase of services, social conflicts over smoking, disliking fellow students). Not at all part of my life 1 2 3 4 5 6 7 8 9 10 Very much part of my life
32.	Social Mistreatment (e.g. social rejection, loneliness, being taken advantage of). Not at all part of my life 1 2 3 4 5 6 7 8 9 10 Very much part of my life
33.	Friendship problems (e.g. conflicts with friends, being let down or disappointed by friends, having your trust betrayed by friends). Not at all part of my life 1 2 3 4 5 6 7 8 9 10 Very much part of my life

Please state how much you agree or disagree with the following statements:

34.	There is a person or people in my life who would provide tangible support for me when I need it (for example: money for tuition or books, use of their car, furniture for a new apartment). Strongly Disagree 1 2 3 4 5 6 7 8 9 10 Strongly Agree
35.	There is a person or people in my life who would provide me with a sense of belonging (for example: I could find someone to go to a movie with me, I often get invited to do things with other people, I regularly hang out with friends). Strongly Disagree 1 2 3 4 5 6 7 8 9 10 Strongly Agree
36.	There is a person or people in my life with whom I would feel perfectly comfortable discussing any problems I might have (for example: difficulties with my social life, getting along with my parents, sexual problems). Strongly Disagree 1 2 3 4 5 6 7 8 9 10 Strongly Agree

References

1. Reid, G.B., Nygren, T.E.: The subjective workload assessment technique: a scaling procedure for measuring mental workload. Adv. Psychol. **52**, 185–218 (1988)
2. Stassen, H.G., Johannsen, G., Moray, N.: Internal representation, internal model, human performance model and mental workload. Automatica **26**(4), 811–820 (1990)
3. De Waard, D.: The measurement of drivers' mental workload. The Traffic Research Centre VSC, University of Groningen (1996)
4. Hart, S.G.: Nasa-task load index (NASA-TLX); 20 years later. In: Human Factors and Ergonomics Society Annual Meeting, vol. 50. Sage Journals (2006)
5. Smith, A.P., Smith, K.: Effects of workload and time of day on performance and mood. In: Megaw, E.D. (ed.) Contemporary Ergonomics, pp. 497–502. Taylor & Francis, London (1988)
6. Evans, M.S., Harborne, D., Smith, A.P.: Developing an objective indicator of fatigue: an alternative mobile version of the psychomotor vigilance task (m-PVT). In: Longo, L., Leva, M.C. (eds.) H-WORKLOAD 2018. CCIS, vol. 1012, pp. 49–71. Springer, Cham (2019). https://doi.org/10.1007/978-3-030-14273-5_4
7. Smith, A.P., Smith, H.N.: Workload, fatigue and performance in the rail industry. In: Longo, L., Leva, M.C. (eds.) H-WORKLOAD 2017. CCIS, vol. 726, pp. 251–263. Springer, Cham (2017). https://doi.org/10.1007/978-3-319-61061-0_17
8. Fan, J., Smith, A.P.: Mental workload and other causes of different types of fatigue in rail staff. In: Longo, L., Leva, M.C. (eds.) H-WORKLOAD 2018. CCIS, vol. 1012, pp. 147–159. Springer, Cham (2019). https://doi.org/10.1007/978-3-030-14273-5_9

9. Cortes Torres, C.C., Sampei, K., Sato, M., Raskar, R., Miki, N.: Workload assessment with eye movement monitoring aided by non-invasive and unobtrusive micro-fabricated optical sensors. In: Adjunct Proceedings of the 28th Annual ACM Symposium on User Interface Software & Technology, pp. 53–54 (2015)
10. Yoshida, Y., Ohwada, H., Mizoguchi, F., Iwasaki, H.: Classifying cognitive load and driving situation with machine learning. Int. J. Mach. Learn. Comput. 4(3), 210–215 (2014)
11. Wilson, G.F., Eggemeier, T.F.: Mental workload measurement. In: Karwowski, W. (ed.) International Encyclopedia of Ergonomics and Human Factors, 2nd edn., Chap. 167, vol. 1. Taylor & Francis, Boca Raton (2006)
12. Young, M.S., Stanton, N.A.: Mental workload. In: Stanton, N.A., Hedge, A., Brookhuis, K., Salas, E., Hendrick, H.W. (eds.) Handbook of Human Factors and Ergonomics Methods, Chap. 39, pp. 1–9. CRC Press, Boca Raton (2004)
13. Young, M.S., Stanton, N.A.: Mental workload: theory, measurement, and application. In: Karwowski, W. (ed.) International Encyclopedia of Ergonomics and Human Factors, vol. 1, 2nd edn, pp. 818–821. Taylor & Francis, Abingdon (2006)
14. Moustafa, K., Luz, S., Longo, L.: Assessment of mental workload: a comparison of machine learning methods and subjective assessment techniques. In: Longo, L., Leva, M.C. (eds.) H-WORKLOAD 2017. CCIS, vol. 726, pp. 30–50. Springer, Cham (2017). https://doi.org/10.1007/978-3-319-61061-0_3
15. Hart, S.G., Staveland, L.E.: Development of NASA-TLX (Task Load Index): Results of Empirical and Theoretical Research. Adv. Psychol. 52(C), 139–183 (1988)
16. Tsang, P.S., Velazquez, V.L.: Diagnosticity and multidimensional subjective work-load ratings. Ergonomics 39(3), 358–381 (1996)
17. Karasek Jr., R.A.: Job demands, job decision latitude, and mental strain: implications for job redesign. Adm. Sci. Q. 24(2), 285–308 (1979)
18. Whitelock, D., Thorpe, M., Galley, R.: Student workload: a case study of its significance, evaluation and management at the Open University. Distance Educ. 36(2), 161–176 (2015)
19. Rummell, C.: An exploratory study of psychology graduate student workload, health, and program satisfaction. Prof. Psychol. Res. Pract. 46(6), 391–399 (2015)
20. Scully, G., Kerr, R.: Student workload and assessment: strategies to manage expectations and inform curriculum development. Acc. Educ. 23(5), 443–466 (2014)
21. Schonfeld, T., Spetman, M.K.: Ethics education for allied health students: an evaluation of student performance. J. Allied Health 36(2), 77–80 (2007)
22. Rozzi, R., De Silvestri, D., Messetti, G.: Pupils' work load and strain in the new Italian primary school (Essere scolari oggi: attivita e fatica. La percezione del carico di lavoro negli insegnanti di scuola elementare dopo la riforma). Eta. Evolutiva. 61, 15–26 (1998). (in Italian)
23. Liu, X.: Measuring teachers' perceptions of grading practices: a cross-cultural perspective. Dissertation Abstracts International Section A: Humanities and Social Sciences, vol. 68(8-A), pp. 3281 (2008)
24. Potts, H.W.: Student experiences of creating and sharing material in online learning. Med. Teach. 33(11), e607–e614 (2011)
25. Kyndt, E., Dochy, F., Struyven, K., Cascallar, E.: The perception of workload and task complexity and its influence on students' approaches to learning: a study in higher education. Eur. J. Psychol. Educ. 26(3), 393–415 (2011)
26. Ruiz-Gallardo, J.-R., Castano, S., Gomez-Alday, J.J., Valdes, A.: Assessing student workload in problem-based learning: relationships among teaching method, student workload and achievement. A case study in natural sciences. Teach. Teach. Educ. 27(3), 619–627 (2011)

27. Williams, G., Pendlebury, H., Thomas, K., Smith, A.P.: The student wellbeing process questionnaire (Student WPQ). Psychology **8**, 1748–1761 (2017)
28. Smith, A.P.: Cognitive fatigue and the well-being and academic attainment of university students. JESBS **24**(2), 1–12 (2018)
29. Williams, G.M., Smith, A.P.: A longitudinal study of the well-being of students using the student well-being questionnaire (WPQ). JESBS **24**(4), 1–6 (2018)
30. Smith, A.P., Smith, H.N., Jelley, T.: Studying away strategies: well-being and quality of university life of international students in the UK. JESBS **26**(4), 1–14 (2018)
31. Omosehin, O., Smith, A.P.: Adding new variables to the Well-being Process Questionnaire (WPQ) – further studies of workers and students. JESBS **28**(3), 1–19 (2019)
32. Alharbi, E., Smith, A.P.: Studying-away strategies: a three-wave longitudinal study of the wellbeing of international students in the United Kingdom. Eur. Educ. Res. **2**(1), 59–77 (2019)
33. Nor, N.I.Z., Smith, A.P.: Psychosocial characteristics, training attitudes and well-being of student: a longitudinal study. JESBS **29**(1), 1–26 (2019)
34. Alharbi, E., Smith, A.P.: Studying away and well-being: a comparison study between international and home students in the UK. IES **12**(6), 1–16 (2019)
35. Olelewe, C.J., Agomuo, E.E.: Effects of B-learning and F2F learning environments on students' achievement in QBASIC programming. Comput. Educ. **103**, 76–86 (2016)
36. Mark, G.M., Smith, A.P.: Stress models: a review and suggested new direction. In: Houdmont, J., Leka, S. (eds.) Occupational Health Psychology: European Perspectives on Research, Education and Practice. EA-OHP Series, vol. 3, pp. 111–144. Nottingham University Press, Nottingham (2008)
37. Mark, G., Smith, A.P.: Effects of occupational stress, job characteristics, coping and attributional style on the mental health and job satisfaction of university employees. Anxiety Stress Copin. **25**(1), 63–78 (2011)
38. Mark, G., Smith, A.P.: Occupational stress, job characteristics, coping and mental health of nurses. Br. J. Health. Psychol. **17**(3), 505–521 (2012)
39. Williams, G.M., Smith, A.P.: Using single-item measures to examine the relationships between work, personality, and well-being in the workplace. Psychol. Spec. Ed. Positive Psychol. **7**, 753–767 (2016)
40. Smith, A.P., Smith, H.N.: An international survey of the wellbeing of employees in the business process outsourcing industry. Psychology **8**, 160–167 (2017)
41. Williams, G., Pendlebury, H., Smith, A.P.: Stress and well-being of nurses: an investigation using the Demands-Resources- Individual Effects (DRIVE) model and Well-being Process Questionnaire (WPQ). Jacobs J. Depression Anxiety **1**, 1–8 (2017)
42. Williams, G., Thomas, K., Smith, A.P.: Stress and well-being of university staff: an investigation using the Demands-Resources- Individual Effects (DRIVE) model and Well-being Process Questionnaire (WPQ). Psychology **8**, 1919–1940 (2017)
43. Jones, M.C., Johnston, D.W.: Distress, stress and coping in first-year student nurses. J. Adv. Nurs. **26**(3), 475–482 (1997)
44. Bayram, N., Bilgel, N.: The prevalence and socio-demographic correlations of depression, anxiety and stress among a group of university students. Soc. Psychiatry Psychiatr. Epidemiol. **43**(8), 667–672 (2008)
45. Dahlin, M., Joneborg, N., Runeson, B.: Stress and depression among medical students: a cross-sectional study. Med. Educ. **39**(6), 594–604 (2005)
46. Tully, A.: Stress, sources of stress and ways of coping among psychiatric nursing students. J. Psychiatr. Ment. Health Nurs. **11**(1), 43–47 (2004)
47. Swickert, R.J., Rosentreter, C.J., Hittner, J.B., Mushrush, J.E.: Extraversion, social support processes, and stress. Personal. Individ. Differ. **32**(5), 877–891 (2002)

48. Kohn, P.M., Lafreniere, K., Gurevich, M.: The inventory of college students' recent life experiences: a decontaminated hassles scale for a special population. J. Behav. Med. **13**(6), 619–630 (1990)
49. Smith, A.P.: Psychosocial predictors of perceptions of academic workload, time pressure, course stress and work efficiency (submitted)
50. Omosehin, O., Smith, A.P.: Do cultural differences play a role in the relationship between time pressure, workload and student well-being? In: Longo, L., Leva, M.C. (eds.) H-WORKLOAD 2019. CCIS, vol. 1107, pp. 186–204. Springer, Cham (2019)

Task Demand Transition Rates of Change Effects on Mental Workload Measures Divergence

Enrique Muñoz-de-Escalona[1](✉), José Juan Cañas[1], and Jair van Nes[2]

[1] Mind, Brain and Behaviour Research Centre,
University of Granada, Granada, Spain
{enriquemef, delagado}@ugr.es
[2] Faculty of Psychology and Neuroscience, University of Maastricht,
Maastricht, The Netherlands
vannes@student.maastrichtuniversity.nl

Abstract. Mental workload is a complex construct that may be indirectly inferred from physiological responses, as well as subjective and performance ratings. Since the three measures should reflect changes in task-load, one would expect convergence, yet divergence between the measures has been reported. A potential explanation could be related to the differential sensitivity of mental workload measures to rates of change in task-load transitions: some measures might be more sensitive to change than the absolute level of task demand. The present study aims to investigate whether this fact could explain certain divergences between mental workload measures. This was tested by manipulating task-load transitions and its rate of change over time during a monitoring experiment and by collecting data on physiological, subjective, and performance measures. The results showed higher pupil size and performance measure sensitivity to abrupt task-load increases: sensitivity to rates of change could partially explain mental workload dissociations and insensitivities between measures.

Keywords: Mental workload · Workload measures · Convergence · Divergence · Dissociations · Insensitivities · Task demand transitions · Rates of change

1 Introduction

The study of mental workload is of crucial importance in many fields such as Emergency Room healthcare or air traffic control (ATC), in which the lives of countless people are at stake and dependent on human performance [1, 2]. Human related accidents address an ongoing problem in social sciences and cognitive ergonomics: how can we minimize and avoid human error?

A high mental workload generally leads to poor performance [3, 4], and extreme cases of overload may result in errors which can in turn lead to fatal accidents. On the other hand, mental underload remains as undesirable as mental overload, likewise leading to poor performance and errors [5]. Automation, which does have its benefits,

L. Longo and M. C. Leva (Eds.): H-WORKLOAD 2019, CCIS 1107, pp. 48–65, 2019.
https://doi.org/10.1007/978-3-030-32423-0_4

is significantly and repeatedly associated with mental underload [6], and can cause a major loss of situation awareness, making it difficult for people to detect flaws and intervene adequately and timely. Therefore, optimizing levels of mental workload is vital to maintaining effective performance, avoiding both overload and underload during task performance. Furthermore, high levels of mental workload can have detrimental effects on people's psychosocial and physical health. High mental workload has been associated with high work-related fatigue, high-stress complaints, and/or burnout [7], as well as high scores on health complaint questionnaires [8]. Therefore, by using mental workload measures, we can learn its limits and dimensions and appropriately discover how to improve performance and minimize human error within organizations or within people's own personal work practices [4]. Task demand is dynamic in many fields such as ATC and aircraft pilots: workers may experience changes in mental workload as, for example, traffic load gets higher or a sudden unexpected storm appears when flying an aircraft, respectively. These changes in taskload may also be gradual or abrupt and affect individuals' mental workload, as well as the way its measures reflect these changes. To better understand and assess the construct of mental workload, the present study aims to investigate whether differential sensitivity to rates of change in task demand transitions would affect the convergence of mental workload measures. However, we will begin by defining mental workload as it in itself is a very poorly defined construct. Furthermore, we will define convergence and divergence phenomena between mental workload measures (associations, dissociations, and insensitivities), as well as the above-mentioned sensitivity to rates of change, to better understand the current mental workload assessment literature. The rest of the paper proceeds as follows. Section 2 outlines related work about general mental workload, focusing on defining and measuring mental workload and on introducing sensitivity to rates of change in task demand transitions. Section 3 describes the design and the methodology followed in conducting the experiment. Section 4 presents the obtained results, while Sect. 5 presents a discussion about our findings, as well as limitations and possible future work. Finally, Sect. 6 concludes the study, summarizing its key findings and implications on the extant body of knowledge.

2 Related Work

2.1 Defining Mental Workload

Mental workload is a complex construct without a clear consensus regarding a definition [9–11]. It has been defined as the product of the immediate demands of the environment and an individual's maximum mental capacity [12, 13], hence mental workload is a multi-factorial construct, which depends not only on demanded task resources but also on those available [14–16]. When the demands of the environment exceed a person's maximum mental capacity, mental overload occurs and performance deteriorates as a consequence of our limited capacity [17]. This is because when at the very limit of our human mental resources, we are unable to reallocate these resources in an adaptable way. On the contrary, when the environment demands too little, such as in work situations that are heavily automated, mental underload occurs and performance similarly deteriorates. Why low environmental demands are detrimental is still poorly understood, however, some have suggested this may be due to shrinkages in our maximum mental capacity in response to environmental demand reduction [5, 18]; this in turn can influence several factors, including vigilance, workload, attention, and situation awareness [6]. For the purpose of our research, mental workload may be considered as the amount of mental effort involved in performing a task [3, 19]. In other words, the amount of mental resources in use during the performance of a task given the demands of the environment. The term task-load refers to environmental demands and it is used to manipulate the amount of experienced mental workload. In general, we say that measures reflecting mental workload are valid when they reflect a change in task-load (the demands of the immediate environment).

2.2 Measuring Mental Workload

The measurement of mental workload is one of the biggest challenges facing psychology and social sciences at present. There is a widespread need to assess cognitive work as, on the one hand, it is fundamental to the development of modern society and, on the other, it has been identified as one of the main causes of work-related accidents. Despite mental workload not being a directly measurable construct, it can be assessed with three types of individual "primary measures" reflecting mental workload [9, 11, 16, 20]: (a) physiological responses (Electroencephalography (other brain imaging techniques, heart-rate variability (HRV), pupil diameter, etc.); (b) perceived or subjective perception of mental workload (questionnaire or scale response); and (c) task performance (response speed and accuracy). According to Hancock (2017), if these three measures of mental workload mean to assess the same construct then they should demonstrate convergence. In other words, if task-load were to increase then we would expect the following associations between task-load changes and primary measures: (a) higher physiological activation responses, (b) higher perceived mental workload, and (c) a decrease in task performance. Thus, there should exist a convergence between these measures of mental workload, given the expected association of task-load with each respective measure. However, current research has demonstrated that this is not always the case: dissociations and insensitivities between mental workload measures have been repeatedly reported [16, 20, 21]. Insensitivities occur when a certain task has distinct levels of task-load, but measures of mental workload fail to reflect a change regardless of task-load levels. For example, when piloting an airplane, pilots may have to deal with dynamic changes within the immediate environment. Task-load may increase in line with the increasing demands of a complex situation, yet our measures reflect static levels of reported mental workload. For instance, task-load may increase due to air turbulence or even a failure in automation, yet physiological measures (such as pupil dilation or HRV) reflect no change. One must note, however, that it is possible for an insensitivity to occur for one measure of mental workload but not for the other. Given the current example, it may be that pilots' physiological measures reflect no change in their actual mental workload, yet they can report an increase (or even decrease) in their perceived mental workload. Furthermore, dissociations occur when we have contradictory results: task-load increases but subjects report a lower perceived mental workload, whereas normally one would expect an increase in perceived mental workload if task-load were to increase. Using our pilot example again, the situation at hand may increase task-load but pilots report (a) lower physiological activation responses (dissociation), (b) higher perceived mental workload (association), and (c) static levels in task performance (insensitivity). There are several factors that might affect the occurrence of dissociation and insensitivities between task-load and primary mental workload measures [20]. One possible explanation might be related to the timescale considered between measures. Muñoz-de-Escalona & Cañas (2018) identified that dissociations and insensitivities may appear at high mental workload peak experiences due to latency differences between measures: subjective measures showed lower levels of latency response than physiological response (pupil size) [16]. Despite this, there are also several other factors that might contribute to the emergence of divergence between mental workload measures, including sensitivity to rates of change [20].

2.3 Task Demand Transitions and Sensitivity to Rates of Change

Task demand transitions research has been very limited and largely focused on its effects in task performance and mental workload perceptions [22]. The current literature has revealed that a change in task-load levels affects mental workload, fatigue, and performance ratings [22–25]. However, the authors could not find any research focusing on how sensitivity to rates of change in task demand transitions affects the convergence and divergence of mental workload measures. Behavioral sciences has repeatedly shown that humans are more sensitive to change rather than the absolute level of a stimulus [20]. If we translate this into the study of mental workload convergence and divergence between measures, it may be possible that there exists a differential sensitivity to task-load rate of change in mental workload measures. Some measures might be more sensitive to change than to the absolute level of task demand, while others might be sensitive only to absolute levels of task demand. These differences would result in dissociations and insensitivities which would ultimately affect convergence between mental workload measures. This study aims to shed light on the effects of sensitivity to rates of change in task demand transitions on the convergence of mental workload measures. In the present study, we manipulated two independent variables: (1) task-load rate of change and (2) task-load change direction, whose effects were tested in a task-battery experiment in which participants were trained and instructed to perform to the best of their abilities. Participants performed the task-battery for 20 min, whilst data on the dependent variables, task performance, pupil diameter, and perceived mental workload were obtained. We hypothesized that there would be higher divergence (dissociations and/or insensitivities) between mental workload measures with abrupt rate of change conditions rather than linear ones.

3 Design and Methodology

3.1 Materials and Instruments

MATB-II Software. Measurements of task performance were collected through the use of the second version of the Multi-Attribute Task Battery (MATB-II), a computer program designed to evaluate operator performance and workload through means of different tasks similar to those carried out by flight crews, with a user-friendly interface as to allow non-pilot participants to utilize it [25]. MATB-II comes with default event files which can easily be altered to adapt to the needs or objectives of an experiment.

The program records events presented to participants, as well as participants' responses. The MATB-II contains the following four tasks: the system monitoring task (SYSMON), the tracking task (TRACK), the communications task (COMM), and the resource management task (RESMAN) (see Fig. 1).

1. The SYSMON task is divided into two sub-tasks: lights and scales. For the lights sub-task, participants are required to respond as fast as possible to a green light that turns off and a red light that turns on, and to turn them back on and off, respectively. For the scale sub-task, participants are required to detect when the lights on four moving scales deviate from their normal position and respond accordingly by clicking on the deviated scale.
2. In the TRACK task, during manual mode, participants are required to keep a circular target in the center of an inner box displayed on the program by using a joystick with their left hand (the dominant hand was needed for the use of the mouse). During automatic mode, the circular target will remain in the inner box by itself.
3. In the COMM task, an audio message is played with a specific call sign and the participant is required to respond by selecting the appropriate radio and adjusting for the correct frequency, but only if the call sign matches their own (call sign: "NASA504"). No response is required of the participant for messages from other call signs.
4. In the RESMAN task, participants are required to maintain the level of fuel in tanks A and B, within ± 500 units of the initial condition of 2500 units each. In order to maintain this objective, participants must transfer fuel from supply tanks to A and B or transfer fuel between the two tanks.

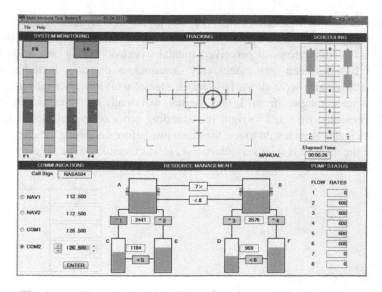

Fig. 1. MATB-II task display. Taken from https://matb.larc.nasa.gov/

Tobii T120 Eyetracker. Pupil diameter measurements were obtained using an infrared-based eye tracker system, the Tobii T120 model marketed by Tobii Video System (see Fig. 2). This system is characterized by its high sampling frequency (120 Hz). This equipment is completely non-intrusive, has no visible eye movement monitoring system, and provides high precision and an excellent head compensatory movement mechanism, which ensures high-quality data collection. In addition, a calibration procedure is completed within seconds, and the freedom of movement it offers participants allows them to act naturally in front of the screen, as though it were an ordinary computer display. Thus, the equipment allows for natural conditions in which to measure eye-tracking data [26].

Fig. 2. Tobii T120 Eyetracker system

Instantaneous Self-assessment Scale. We employed an easy and intuitive instant subjective workload scale called instantaneous self-assessment (ISA), which provides momentary subjective ratings of perceived mental workload during task performance (see Fig. 3). ISA has been used extensively in numerous domains, including during ATC tasks. Participants write down how much mental workload they currently experience on a scale ranging from 1 (no mental workload) to 5 (maximum mental workload), presented from left to right in ascending order of mental workload experienced. Participants were taught to use the scale just before beginning the experimental stage. While the method is relatively obtrusive, it was considered the least intrusive of the available online workload assessment techniques [27, 28].

Participante: Edad:

ISA SCALE

Responda según las siguientes opciones:

1 (Escasa de carga), 2 (Algo de carga), 3 (Carga moderada), 4 (Bastante carga), 5 (Máxima carga)

2.5 MIN————————————————————— 1 2 3 4 5

5 MIN————————————————————————— 1 2 3 4 5

7.5 MIN———————————————————————— 1 2 3 4 5

10 MIN———————————————————————— 1 2 3 4 5

12.5 MIN—————————————————————— 1 2 3 4 5

15 MIN———————————————————————— 1 2 3 4 5

17.5 MIN—————————————————————— 1 2 3 4 5

20 MIN———————————————————————— 1 2 3 4 5

Fig. 3. Instantaneous self-assessment scale

3.2 Participants

Fifty-six psychology students from the University of Granada participated in the study. Participants' ages ranged from 18 to 30, with an average of 22.7 and a standard deviation of 4. A total of 39 women and 17 men participated. It should be noted that there is a greater number of female participants due to the fact that psychology students at the University of Granada are mostly women. Recruitment was achieved through the dispersion of posters and flyers around the university, as well as an advertisement for the study on the university's online platform for experiments (http://experimentos. psiexpugr.es/). The requirements for participation included (1) not being familiar with the MATB-II program, (2) Spanish as a native language, and (3) visual acuity or correction of visual impairment with contact lenses, as glasses impair the utilized eye-tracking device from collecting data. Participants' participation was rewarded with two experimental vouchers for which they received extra credit.

3.3 Procedure

1. **Training stage:** training took place for no longer than 30 min. The objective of this stage was for participants to familiarize themselves with the program so that they could carry out the tasks securely during the data collection stage. The procedure was conducted as follows: upon entering the lab and after filling out the informed consent form, the participant was instructed to read the MATB-II instruction manual and inform the researcher once they had finished. The researcher then sat down with the participant to allow for questions and resolve any doubts on how to use the program. Afterward, on a computer monitor, participants were presented each MATB-II task separately and were first given a demonstration as to how to execute

the task and given time to perform the task themselves. The participants were always free to consult the manual and ask the researcher questions during the training stage in case of doubts or uncertainties. Once the participants had completed all four tasks and resolved all doubts, they were ready for the data collection stage, which followed immediately afterwards. During the training stage, participants could work in one of three different rooms equipped for training with the MATB-II software, and no special attention to room conditions was needed.

2. **Data collection stage:** the data collection stage lasted approximately 20 min and involved participants completing 1 of the 4 randomly assigned experimental conditions, while task performance, perceived mental workload, and pupil diameter were recorded. The participants were instructed to fill in the ISA scale every 2 and a half minutes when a scheduled alarm sounded. Prior to the start of the task-battery, the eye-tracker system was calibrated, and the participants were told to keep head and body movements to a minimum. During the data collection stage, standardizing room conditions was essential. Thus, the testing rooms were temperature controlled to 21 °C, and lighting conditions (the main extraneous variable in pupil diameter measurement) were kept constant with artificial lighting; there was no natural light in the rooms. Moreover, participants always sat in the same place, a comfortable chair spaced 60 cm from the eye-tracker system.

This study was carried out in accordance with the recommendations of the local ethical guidelines of the committee of the University of Granada institution: Comité de Ética de Investigación Humana. The protocol was approved by the Comité de Ética de Investigación Humana under the code: 779/CEIH/2019. All subjects gave written informed consent in accordance with the Declaration of Helsinki.

3.4 Variables

Independent Variables:
In the present study we manipulated 2 independent variables:

- *Task-load rate of change:* this is the intensity in which task demand changes over time. We manipulated the task-load rate of change by modifying the number and the combination of active tasks that participants had to perform over time; this occurred during the data collection stage. We established 2 levels: (1) the variable rate of change and (2) the linear rate of change. Possible task combinations are illustrated in Table 1.

Table 1. Possible task combination in MATB-II software

Task Combination	MATB-II Active tasks
1	SYSMON
2	SYSMON + TRACK
3	SYSMON + TRACK + COMM
4	SYSMON + COMM + RESMAN
5	SYSMON + TRACK + COMM + RESSMAN

- *Task-load change direction*: this is the direction in which task-load changes over time. Since task demand transitions can occur in two directions, we manipulated this variable on 2 levels: (1) increasing task-load change and (2) decreasing task-load change.

As a result of the manipulation of these two variables, we obtained 4 experimental conditions (see Fig. 4), namely:

1. **Condition: Increasing task-load with a variable rate of change.** Task-load increased every 5 min, but with a variable rate of change. Participants had to perform 1, 2, 4, and 5 sets of task combinations: there is an initial rate of change from task combination 1 to 2, then there is an abrupt increase in the rate of change from task combination 2 to 4 and then a decrease in the rate of change from task combination 4 to 5.
2. **Condition: Decreasing task-load with a variable rate of change.** Participants performed the same set of tasks from condition (1) but in descending order, resulting in 5, 4, 2, and 1 task combinations.
3. **Condition: Increasing task-load with a linear rate of change.** Task-load increased every 5 min with a linear rate of change. Participants had to perform task combinations 1, 2, 3, and 5.
4. **Condition: Decreasing task-load with a linear rate of change.** Participants performed the same set of tasks from condition (3) but in descending order, resulting in task combinations 5, 3, 2, and 1.

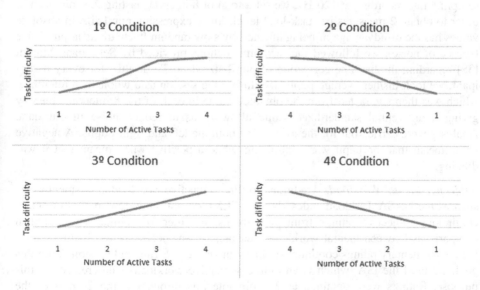

Fig. 4. Task-load rate of change evolution for experimental conditions.

Note that the difference between the increasing variable and linear rate of change conditions demonstrates that the former condition involves a sharp increase in task-demands from task combination 2 to 4. The reason for this lies with the elimination of an already practiced task (TRACKING) and the addition of 2 new non-practiced tasks (COMM & RESMAN), whereas the latter conditions involve a linear increase in task-demands from task combination 2 to 3 since only a single new task is added (COMM) and vice-versa regarding a decreasing variable and a linear rate of change conditions.

Dependent Variables:

Performance. MATB-II provides us with many indicators of participants' performance: e.g. root mean square deviation (RMSD) for the TRACK task, number of correct and incorrect responses for the SYSMON and COMM tasks, and the arithmetic mean of tanks "A-2500" and "B-2500" in absolute values for the RESMAN task. However, for the purposes of this experiment we will only consider the SYSMON performance indicator, as it is the only task present during all 4 of the task-load levels in the 4 experimental conditions, allowing us to compare participants' performances between conditions. The SYSMON performance indicator will be considered as the number of correct responses divided into the number of possible responses. The result is a number between 1 (best possible performance) and 0 (worse possible performance).

Pupil Size. Mental workload can be reflected by several physiological indexes such as EEG, HVR, and several ocular metrics. We decided to use pupil diameter as our physiological mental workload indicator, as it effectively reflects mental workload [29–37] and minimize intrusiveness. While our eye-tracking system allows continuous sampling rate recording at 120 Hz, we set a total of 8 intervals lasting 2.5 min each in order to obtain 2 measures per task-load level. Since expressing pupil size in absolute values has the disadvantage of being affected by slow random fluctuations in pupil size (source of noise), we followed the recommendations provided by Sebastiaan Mathôt [38] regarding the baseline correction of pupil-size data. To do this, for every participant, we took his/her average pupil size during the session as a whole as a reference, which was then subtracted from the obtained value in each of the 8 intervals, thereby giving a differential standardized value allowing us to reduce noise in our data. Analyses were carried out for the average of both the left and right pupils. A negative value meant that the pupil was contracting while a positive value meant that it was dilating.

Subjective Mental Workload. Traditional offline subjective workload assessment tools, such as the NASA Taskload Index (NASA-TLX), do not allow researchers to obtain continuous subjective ratings from participants. In order to facilitate and establish comparisons between mental workload measures, it was necessary to obtain the subjective momentary ratings continuously throughout the experimental session. With this goal, we used the ISA, which is an online subjective workload scale created for this purpose. Ratings were obtained at 2.5-min intervals throughout the 20 min of the experimental stage, obtaining a total of 8 subjective mental workload ratings (2 measures per task demand level).

Synchronization of Measures. Performance, pupil size, and subjective measures were obtained continuously throughout the experimental session. Synchronization between measures was simple, as the eyetracker and MATB-II performance log files began to record data simultaneously at the start of the experimental session. The scheduled alarm (every 2.5 min) was also synchronized by the experimenter, as it was simultaneously activated with the MATB-II software. This would also allow the ISA scale to be synchronized with the performance and pupil size measures.

4   Results

We used three one-way, within-subjects ANOVA to analyze the obtained results, one for each mental workload measurement. First, the analyses of participants' performance showed that our task-load level manipulation was successful. The ANOVA analyses identified a very significant main task-load level effect $F(3,156) = 74.34$, MSe = .005, p < .01, which reflects that participants' performances decreased as task demand increased. The main effect of the experimental condition was found to be significant F $(3,52) = 3.24$, MSe = .02, p < .01. Moreover, an interaction effect of the task-load level x experimental condition was also found to be significant $F(9,156) = 6.14$, MSe = .005, p < .01. This demonstrates that performance variations evolve differently throughout the different task-load levels depending on the considered experimental condition: there is a higher decrease in performance in the variable rate of change conditions compared to the linear rate of change conditions from task-load levels 2 to 3 (see Fig. 5).

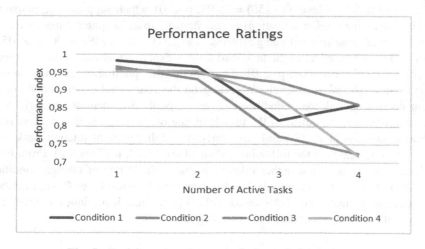

Fig. 5. Participants' performance during task development.

Regarding subjective perceptions, a linear increase in participants' subjective mental workload ratings occurs between task-load levels in every experimental condition: in the ANOVA, the main effect of task-load levels turned out to be very significant indeed, F $(3,156) = 358.9$; MSe = .240, p < .01; whereas the main effect of the experimental condition F $(3,52) = 1.09$; MSe = .84, p > .05 and the interaction of

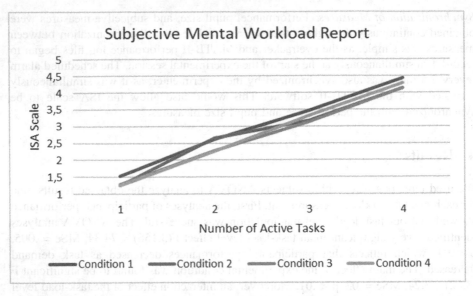

Fig. 6. Participants' subjective mental workload ratings during task development.

the task-load level x experimental condition F (9,156) = .72; MSe = .240, p > .05 were not significant (see Fig. 6).

For pupil size, our physiological measurement, we also found a very significant main effect of task-load level F(3,156) = 70.94, p < .01 which supported participants' pupil size increasing (higher activation) as task demand rose. Despite the main effect of the experimental condition not being significant F(3,52) = 1.03, MSe = .00, p > .05, a significant interaction effect of the task-load level x experimental condition was found F(9,156) = 4.93; MSe = .008, p < .01. This implies that pupil size variation evolves differently through task-load levels depending on the considered experimental condition: in decreasing task-load conditions (2 & 4), pupil size increases linearly with higher task demand, regardless of the task-load rate of change; whereas in increasing task-load conditions (1 & 3), pupil size starts at a higher state of dilation (task-load level 1), then decreases in the following task-load level (task-load level 2). From task-load level 2 to 3, there is a higher dilation in the variable rate of change condition (1) regarding the linear condition (3). Finally, from task-load level 3 to 4, pupil dilation decreases in the variable rate of change condition (1), whereas it continues increasing in the linear rate of change condition (3) (see Fig. 7).

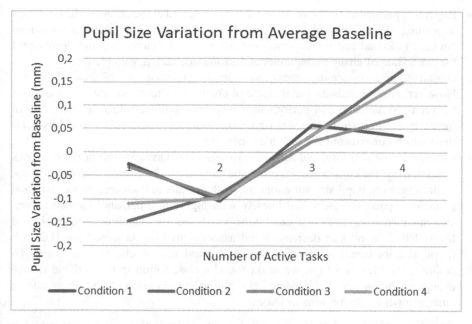

Fig. 7. Participants' pupil size variation ratings from the average baseline during task development.

5 Discussion

To synthesize our data, we can see how subjective ratings increased linearly as task-load rose in every experimental condition, whereas the performance and physiological measures reacted differently depending on the experimental condition in question. Performance linearly decreased as task-load increased in both (increasing and decreasing) lineal rate of change conditions (3 & 4), as well as in the decreasing variable rate of change condition (2). Conversely, in the increasing variable rate of change condition (1), performance decreased until task-load level 3 and then improved again from task-load levels 3 to 4. Pupil size increased linearly in both decreasing task-load conditions (2 & 4) as task demand increased; while in both increasing task-load conditions there was a decrease in pupil size from task-load levels 1 to 2, followed by an increase from task-load levels 2 to 3. Finally, from task-load levels 3 to 4, pupil size continued increasing in the linear rate of change condition but decreased in its variable rate of change counterpart.

Hence, according to our data and in terms of association, dissociation, and insensitivities:

- Regarding *subjective response*, we found associations between task-load and sub-jective perceptions in every experimental condition: there is a direct mapping between mental workload subjective ratings and task demand. We can observe how subjective ratings have not been influenced either by task-load rate of change or task-load change direction.

- Regarding *performance response*, our results presented associations or dissociations depending on the considered experimental condition: we found associations between task-load and performance responses in both (increasing and decreasing) the linear rate of change conditions and in the decreasing variable rate of change condition; performance became more linearly impaired as task-load increased. However, in the increasing variable rate of change condition, we found associations between task demand and performance responses until task-load level 3, but an improvement in performance was observed from task-load levels 3 to 4 (dissociation) which contradicted the task-load increase.
- Concerning our *physiological response*, we also found associations or dissociations depending on the considered experimental condition: our results discovered associations between pupil size variations and task demand in both decreasing task-load conditions: pupil size increased linearly with higher task demand. However, considering increasing task-load conditions we found dissociations from task-load levels 1 to 2 (pupil size decreases) and associations from task-load levels 2 to 3 (pupil size increases) in both variable and lineal rate of change conditions. To contrast, from levels 3 to 4, we again found a dissociation in the variable rate of change condition (pupil size decreases), but an association with the linear rate of change condition (pupil size increases).

Therefore, this data partially confirmed our hypothesis that there exists a higher divergence between mental workload measures with variable rate of change conditions: we found performance dissociation only in the increasing task-load with a variable rate of change condition (condition 1). Furthermore, we found physiological dissociations in both variable rate of change conditions from task-levels 1 to 2, but from task-load levels 3 to 4 we found a dissociation only in the increasing variable rate of change condition. Hence, taking into account our results, we could say that there exists a higher divergence between mental workload measures with a variable rate of change condition, particularly in increasing task-load variable rate of change conditions. These results could be explained under the explanation provided by Hancock (2017, page 12) in which he claims that ...[one of the more well-established principles that we do have in the behavioural sciences is that human frequently prove more sensitive to change rather than the absolute level of a stimulus array]... [20]. In other words, some mental workload reflections could be more sensitive to change rather than the absolute level of task demand, whereas others might be more sensitive to the absolute levels of task demand. Therefore, by analyzing our results, subjective perception ratings appeared to be less sensitive to task demand rate of change: they linearly increased in every experimental condition showing no statistical differences among them. Nevertheless, performance measures appeared to be more sensitive to an abrupt increase in task-load level, as the worst performance peak in the increasing variable rate of change condition was achieved in the abrupt transition from task-load levels 2 to 3, even though task demand was higher in task-load level 4. Moreover, physiological measures also appeared to be more sensitive to abrupt changes in task demand only in the increasing task-load condition. Similarly to what happened with performance measures, a higher pupil size peak was achieved in the abrupt transition from task-load levels 2 to 3, despite task-load being higher in task-load level 4. As pupil dilation reflects activation (among other factors which were controlled), when there is an abrupt increase in task demand, we seem to overreact in order to prepare ourselves to face environmental threats but, in line with resources theories, due to the fact that mental resources are limited and can be depleted, when an abrupt increase is followed by a soft increase in task-load, our organism detects that there is no need to continue activating and it deactivates in order to save resources. On the other hand, higher pupil size in task-load level 1 for both increasing task-load conditions (compared to task-load level 2 in the same condition, and compared to task-load level 1 in the decreasing task-load conditions) could be explained by the fact that participants' activation were higher at the beginning of the experimental session because of the natural nervousness experienced by participants. While in both the decrement task-load conditions, this nervousness activation is added to the activation produced by task-demand level 4, as is reflected in our data: pupil dilation for decreasing task-load conditions were higher in task-load level 4 than for increasing task-load conditions. These findings should be viewed in light of some study limitations. Although we were able to overcome other studies' limitations, such as the examination of a single direction task demand transition [39], because of the high number of possible combinations, it was not suitable to analyse other interesting experimental conditions (as in for example, low-high-low/high-low-high transitions or changing the intervals in which the abrupt change in task-load demand occurs: beginning, middle, and end of the scenario). Moreover, we think it would be highly interesting to introduce other physiological measures in this study, such as EEG and/or HRV. We must bear in mind that divergences have been found not only between the three primary mental workload measures (performance, physiological, and subjective), but within different indexes of each primary indicator. Lastly, this study has been conducted with students under simulated conditions and we think that it would be appropriate to validate these findings under real conditions in order to improve ecological validity.

Further research is needed to untangle mental workload divergence between measures. Future research could address the aforementioned methodological limitations. For example, it would be interesting to analyse how sensitivity to the rate of change effects varies depending on when the abrupt change takes place or how low-high-low/high-low-high affects mental workload measures' convergence.

6   Conclusions

Mental workload is a complex construct which can be measured by its three primary measures: performance, physiological, and subjective. Despite expecting to find convergence between them as they reflect the same construct, dissociations and insensitivities have been repeatedly reported in the literature. A potential explanation for these divergences could be related to the differential sensitivity of mental workload measures to task-load transitions rate of change: some measures might be more sensitive to change than the absolute level of task demand, while others might be more sensitive to absolute levels of task demand. These differences would result in dissociations and insensitivities, which would ultimately affect convergence between mental workload measures. Our results suggest that dissociations in performance and physiological pupil size measures may appear after an abrupt change takes place, albeit mostly during increasing task-load conditions. However, subjective ratings may not be affected by the task-load rate of change but by the absolute level of task demand. In other words, our results partially confirmed our hypothesis, as we found higher divergence (dissociations and/or insensitivities) between mental workload measures with abrupt rates of change but only during the increasing task-load condition. An important implication of this finding is that we should give more weight to one of the mental workload reflections depending on environmental rate of change demands. In other words, if task-load transitions are linear, then we could rely on every primary mental workload measure; but when there is an abrupt task-demand transition from low to high mental workload, we may prefer to rely on subjective ratings rather than physiological or performance measures, as the subsequent decrement in physiological activation (saving resources) would not necessarily mean a decrement in an operator experienced mental workload.

References

1. Pape, A.M., Wiegmann, D.A., Shappell, S.A.: Air traffic control (ATC) related accidents and incidents: A human factors analysis (2001)
2. Reuters: What We Know About the Deadly Aeroflot Superjet Crash Landing, 6th May 2019. https://www.themoscowtimes.com/2019/05/06/what-we-know-about-the-deadly-aeroflot-superjet-crash-landing-a65495
3. Byrne, A.J., Sellen, A.J., Jones, J.G.: Errors on anaesthetic record charts as a measure of anaesthetic performance during simulated critical incidents. Br. J. Anaesth. **80**(1), 58–62 (1998)
4. Byrne, A.: Measurement of mental workload in clinical medicine: a review study. Anesth. Pain Med. **1**(2), 90 (2011). https://doi.org/10.5812/kowsar.22287523.2045
5. Young, M.S., Stanton, N.A.: Attention and automation: new perspectives on mental underload and performance. Theor. Issues Ergon. Sci. **3**(2), 178–194 (2002). https://doi-org.ezproxy.ub.unimaas.nl/10.1080/14639220210123789
6. Endsley, M.R.: From here to autonomy: lessons learned from human–automation research. Hum. Factors **59**(1), 5–27 (2017). https://doi-org.ezproxy.ub.unimaas.nl/10.1177%2F0018720816681350
7. Zoer, I., Ruitenburg, M.M., Botje, D., Frings-Dresen, M.H.W., Sluiter, J.K.: The associations between psychosocial workload and mental health complaints in different age groups. Ergonomics **54**(10), 943–952 (2011). https://doi.org/10.1080/00140139.2011.606920
8. Kawada, T., Ooya, M.: Workload and health complaints in overtime workers: a survey. Arch. Med. Res. **36**(5), 594–597 (2005). https://doi-org.ezproxy.ub.unimaas.nl/10.1016/j.arcmed.2005.03.048
9. Cain, B.: A review of the mental workload literature. Defence Research and Development Toronto (Canada) (2007)
10. Meshkati, N., Hancock, P.A. (eds.): Human Mental Workload, vol. 52. Elsevier, Amsterdam (1988)
11. Moray, N. (ed.): Mental Workload: Its Theory and Measurement, vol. 8. Springer, New York (1979). https://doi.org/10.1007/978-1-4757-0884-4

12. Kantowitz, B.H.: Attention and mental workload. In: Proceedings of the Human Factors and Ergonomics Society Annual Meeting, vol. 44, no. 21, pp. 3–456, July 2000. https://doi-org. ezproxy.ub.unimaas.nl/10.1177%2F154193120004402121

13. Wickens, C.D.: Multiple resources and mental workload. Hum. Factors **50**(3), 449–455 (2008). https://doi.org/10.1518/001872008X288394

14. Wickens, C.D.: Multiple resources and performance prediction. Theor. Issues Ergon. Sci. **3**(2), 159–177 (2002). https://doi.org/10.1518/001872008x288394. 2008 50:449

15. Munoz-de-Escalon, E., Canas, J.: Online measuring of available resources. In: H-Workload 2017: The First International Symposium on Human Mental Workload, Dublin Institute of Technology, Dublin, Ireland, 28–30 June (2017). https://doi.org/10.21427/d7dk96

16. Muñoz-de-Escalona, E., Cañas, J.J.: Latency differences between mental workload measures in detecting workload changes. In: Longo, L., Leva, M.C. (eds.) H-WORKLOAD 2018. CCIS, vol. 1012, pp. 131–146. Springer, Cham (2019). https://doi.org/10.1007/978-3-030-14273-5_8

17. Durantin, G., Gagnon, J.F., Tremblay, S., Dehais, F.: Using near infrared spectroscopy and heart rate variability to detect mental overload. Behav. Brain Res. **259**, 16–23 (2014). https://doi.org/10.1016/j.bbr.2013.10.042

18. Young, M.S., Stanton, N.A.: Malleable attentional resources theory: a new explanation for the effects of mental underload on performance. Hum. Factors **44**(3), 365–375 (2002). https://doi.org/10.1518/0018720024497709

19. Byrne, A.J., et al.: Novel method of measuring the mental workload of anaesthetists during clinical practice. Br. J. Anaesth. **105**(6), 767–771. https://doi.org/10.1093/bja/aep268

20. Hancock, P.A.: Whither workload? Mapping a path for its future development. In: Longo, L., Leva, M.Chiara (eds.) H-WORKLOAD 2017. CCIS, vol. 726, pp. 3–17. Springer, Cham (2017). https://doi.org/10.1007/978-3-319-61061-0_1

21. Yeh, Y.Y., Wickens, C.D.: Dissociation of performance and subjective measures of workload. Hum. Factors **30**(1), 111–120 (1988). https://doi-org.ezproxy.ub.unimaas.nl/10.1177%2F001872088803000110

22. Edwards, T., Martin, L., Bienert, N., Mercer, J.: The relationship between workload and performance in air traffic control: exploring the influence of levels of automation and variation in task demand. In: Longo, L., Leva, M. (eds.) H-WORKLOAD 2017. CCIS, vol. 726, pp. 120–139. Springer, Cham (2017). https://doi.org/10.1007/978-3-319-61061-0_8

23. Helton, W.S., Shaw, T., Warm, J.S., Matthews, G., Hancock, P.: Effects of warned and unwarned demand transitions on vigilance performance and stress. Anxiety Stress Coping **21**, 173–184 (2008)

24. Cox-Fuenzalida, L.E.: Effect of workload history on task performance. Hum. Factors **49**, 277–291 (2007)

25. Santiago-Espada, Y., Myer, R.R., Latorella, K.A., Comstock Jr., J.R.: The multi-attribute task battery II (MATB-II) software for human performance and workload research: a user's guide (2011)

26. Lee, J., Ahn, J.H.: Attention to banner ads and their effectiveness: an eye-tracking approach. Int. J. Electron. Commer. **17**(1), 119–137 (2012). https://doi.org/10.2753/JEC1086-4415170105

27. Brennan, S.D.: An experimental report on rating scale descriptor sets for the instantaneous self assessment (ISA) recorder. DRA Technical Memorandum (CAD5) 92017, DRA Maritime Command and Control Division, Portsmouth (1992)

28. Jordan, C.S.: Experimental study of the effect of an instantaneous self assessment workload recorder on task performance. DRA Technical Memorandum (CAD5) 92011. DRA Maritime Command Control Division, Portsmouth (1992)

29. Matthews, G., Middleton, W., Gilmartin, B., Bullimore, M.A.: Pupillary diameter and cognitive and cognitive load. J. Psychophysiol. **5**, 265–271 (1991)
30. Backs, R.W., Walrath, L.C.: Eye movement and pupillary response indices of mental workload during visual search of symbolic displays. Appl. Ergon. **23**, 243–254 (1992). https://doi.org/10.1016/0003-6870(92)90152-l
31. Hyönä, J., Tommola, J., Alaja, A.: Pupil dilation as a measure of processing load in simultaneous interpreting and other language tasks. Q. J. Exp. Psychol. **48**, 598–612 (1995). https://doi.org/10.1080/14640749508401407
32. Granholm, E., Asarnow, R.F., Sarkin, A.J., Dykes, K.L.: Pupillary responses index cognitive resource limitations. Psychophysiology **33**, 457–461 (1996). https://doi.org/10.1111/j.1469-8986.1996.tb01071.x
33. Iqbal, S.T., Zheng, X.S., Bailey, B.P.: Task evoked pupillary response to mental workload in human-computer interaction. In: Proceedings of the ACM Conference on Human Factors in Computing Systems, pp. 1477–1480. ACM, New York (2004). https://doi.org/10.1145/985921.986094
34. Verney, S.P., Granholm, E., Marshall, S.P.: Pupillary responses on the visual backward masking task reflect general cognitive ability. Int. J. Psychophysiol. **52**, 23–36 (2004). https://doi.org/10.1016/j.ijpsycho.2003.12.003
35. Porter, G., Troscianko, T., Gilchrist, I.D.: Effort during visual search and counting: insights from pupillometry. Q. J. Exp. Psychol. **60**, 211–229 (2007). https://doi.org/10.1080/17470210600673818
36. Priviter, C.M., Renninger, L.W., Carney, T., Klein, S., Aguilar, M.: Pupil dilation during visual target detection. J. Vision **10**, 1–14 (2010). https://doi.org/10.1167/10.10.3
37. Reiner, M., Gelfeld, T.M.: Estimating mental workload through event-related fluctuations of pupil area during a task in a virtual world. Int. J. Psychophysiol. **93**(1), 38–44 (2014)
38. Mathôt, S., Fabius, J., Van Heusden, E., Van der Stigchel, S.: Safe and sensible preprocessing and baseline correction of pupil-size data. Behav. Res. Methods **50**(1), 94–106 (2018). https://doi.org/10.3758/s13428-017-1007-2
39. Morgan, J.F., Hancock, P.A.: The effect of prior task loading on mental workload: an example of hysteresis in driving. Hum. Factors **53**(1), 75–86 (2011)

Validation of a Physiological Approach to Measure Cognitive Workload: CAPT PICARD

Bethany Bracken[1]([⊠]), Calvin Leather[1], E. Vincent Cross II[2],
Jerri Stephenson[3], Maya Greene[4], Jeff Lancaster[5], Brandin Munson[6],
and Kritina Holden[2]

[1] Charles River Analytics, 625 Mount Auburn St., Cambridge, MA 02138, USA
{bbracken, cleather}@cra.com
[2] Leidos, Reston, USA
crossev@gmail.com, kritina.l.holden@nasa.gov
[3] NASA Johnson Space Center, Houston, USA
jerri.l.stephenson@nasa.gov
[4] KBRwyle, Houston, USA
maya.r.greene@gmail.com
[5] Honeywell, Charlotte, USA
jlancast5@yahoo.com
[6] GeoLogics, Alexandria, USA
brandin.a.munson@nasa.gov

Abstract. This study validated a physiological workload tool during an investigation of information integration impacts on performance of a NASA electronic procedures task. It was hypothesized that as the level of integration of system and procedural information increased, situation awareness (SA) and usability would increase, and cognitive workload would decrease. To allow quantitative, continuous, and objective assessment of cognitive workload, Charles River Analytics designed a system for Cognitive Assessment and Prediction to Promote Individualized Capability Augmentation and Reduce Decrement (CAPT PICARD). This real-time workload tool was validated against the Bedford workload rating scale in this NASA operational study. The overall results for SA and cognitive workload were mixed; however, the CAPT PICARD workload and eye tracking measures (both real-time physiological measures), showed congruous results. The value of workload assessment during developmental testing of a new system is that it allows for early identification of features and designs that result in high workload. When issues are identified early, redesigns are more feasible and less costly. Real-time workload data could provide the inputs needed to drive future adaptive displays, (e.g., if an astronaut is experiencing high cognitive workload, this data could cue an option for simplified displays). A future vision for real-time, unobtrusive measurement of workload is also to use it in an operational spaceflight environment. Therefore, it is critically important to continue to test and mature tools for unobtrusive measures of human performance like CAPT PICARD.

E. Vincent Cross II—Formerly Leidos M. Greene—Formerly KBRwyle J. Lancaster—Formerly Honeywell B. Munson—Formerly GeoLogics.

L. Longo and M. C. Leva (Eds.): H-WORKLOAD 2019, CCIS 1107, pp. 66–84, 2019.
https://doi.org/10.1007/978-3-030-32423-0_5

Keywords: Cognitive workload · Situation awareness · SPAM · Bedford workload scale · Functional near infrared spectroscopy (fNIRS) · Eye tracking

1 Introduction

NASA is continually developing new technologies to improve astronaut effectiveness. However, if developed incorrectly, these technologies risk being overly burdensome, thereby negating any performance improvement, and possibly even hindering crew and endangering their mission. Current methods of assessing cognitive workload are subjective, and must either interrupt task performance or be delivered posthoc (e.g., the NASA TLX, the Bedford Workload Scale). An unobtrusive system to measure astronaut workload during testing and evaluation (T&E), when tools and systems are being evaluated, would allow engineers to accurately evaluate the cognitive demands of these tools, and the effects they will have on task performance and accuracy. Having this type of measurement capability in the operational space environment would be a prerequisite for future adaptive systems, but that is a longer-term goal. To address the current need, Charles River Analytics developed a system for Cognitive Assessment and Prediction to Promote Individualized Capability Augmentation and Reduce Decrement (CAPT PICARD). This real-time physiological workload tool was tested as part of a NASA study on the effects of electronic procedure/display integration on situation awareness (SA) and workload.

1.1 Quantitative, Objective, Continuous Cognitive Workload Assessment

The most effective means of measuring mental states such as cognitive workload is by measuring brain activity directly. Multiple neurophysiological sensors are useful for this task. For example, electroencephalography (EEG) is useful for assessing cognitive workload; however EEG sensors are highly sensitive to motion artifacts [1], require significant training to learn how to set up and use, have output measures that are difficult to translate into a form that is easily understandable (e.g., event-related potentials), and typically require post-hoc processing, which prevents real-time assessment. Functional Near-Infrared Spectroscopy (fNIRS) is a direct and quantitative method to measure ongoing changes in brain blood oxygenation (HbO) in response to a subject's evolving cognitive state, mental workload, or executed motor actions [2, 3] that has only recently received significant attention from researchers and product developers. CAPT PICARD assesses cognitive workload using fNIRS, which has been well-established in the lab environment [2–5]. Increased workload corresponds with increases in prefrontal oxygenated blood volume (HbO), which is correlated with increased task engagement. Once the task becomes too difficult, HbO decreases, as does task engagement and performance [4]. However, fNIRS sensors useful for assessing HbO during normal activities in real-world environments are only recently emerging, and standard brain activity sensors are large (e.g., full-head), expensive (~ $10K+), and require heavy equipment (e.g., batteries, laptops). These solutions are not feasible for use in space, where there are many constraints on mass, power, and crew mobility. To address this gap, Charles River developed a custom fNIRS sensor (patent pending, and now undergoing active commercial sale). The first-generation sensor is the *fNIRS Pioneer*™, which can accurately assess HbO and cardiac information during activities with low physical activity (e.g., naturalistic movement within a confined environment such as a mock cockpit). The Pioneer is shown in Fig. 1. It is roughly the size of a business card; has a unit cost of ~$500; fits comfortably into standard headwear—including baseball caps—without modifications; weighs 2 oz; and runs for at least 24 h continuously after a full charge. This more portable fNIRS sensor is used to non-invasively assess changes in the HbO in the brain, which provides a robust, accurate, and real-time assessment of cognitive workload [6, 7].

Fig. 1. Our commercial (TRL 8), ruggedized, field-ready fNIRS Pioneer™ (left)

As sensor data is collected, it is processed and modeled into data indicative of cognitive state. Collection of data in testing and evaluation labs, such as those at NASA that often require naturalistic participant motion (e.g., looking back and forth between multiple display screens and interacting with a console), can result in noisy and missing data. To address this, Charles River developed its Sherlock™ software product to identify and validate the links between participant behaviors during performance of tasks and the cognitive states they reflect. We built a *Data Fusion Engine* into Sherlock to handle messy data characteristic of real-world tasks. During real-time data ingestion, we apply custom noise rejection and signal recovery methods during motion, allowing detection and removal of impulse and other motion-related artifacts while retaining relevant information, including physiological components such as the cardiac frequency band and task-related HbO changes. Charles River has also optimized the filtering algorithm to properly handle datasets that start with a period of noisy/unprocessable data (allowing recalibration of artifact reduction and featurization for subsequent periods with different levels of noise). We use robust metrics to assess signal and noise characteristics (median average deviation rather than standard deviation) to filter out meaningless outliers. The Data Fusion Engine also applies a combination of advanced motion correction algorithms including wavelet filtering, movement artifact removal algorithm (MARA), and acceleration-based movement artifact reduction algorithm (AMARA) [8–10]. Charles River has now developed and evaluated wavelet, MARA, and AMARA filtering techniques alone and in combination within Sherlock, applied these techniques to real-world, noisy data, and validated this approach to filter out large motion-related noise events. We filtered the data to the point that we were able to salvage data from even high-motion conditions. This has substantially improved the stability of the physiological measures produced by the sensors. Sherlock also includes a *Data Interpretation Engine* to interpret human state (e.g., cognitive workload) during ecologically valid tasks, both posthoc and in real time. Sherlock enables the simultaneous use of multiple statistical (e.g., weighted averaging [11]) and probabilistic modeling techniques (e.g., Bayesian networks (BNs), dynamic BNs (DBNs) [12], causal influence models (CIMs) [13]), and machine-learning algorithms to construct general purpose models that identify causal links between human states (such as cognitive workload) and the physiological and neurophysiological signals underlying these states.

1.2   Electronic Aerospace Procedures

Aerospace procedures have evolved from memory tools into support tools that help to ensure adequate, if not improved, cognitive burden and SA [14]. Different types of procedures and checklists can be found throughout aerospace, often stratified between normal/nominal, abnormal/off-nominal, and emergency procedures. Starting in the late 1990s, what had been a largely paper-based procedural regime evolved into an electronic one that is not only much less physically cumbersome (e.g., carrying/managing one or more paper procedure manuals), but one that also helps to overcome several design weaknesses associated with paper, including: the lack of a pointer to the current item; an inability to mark skipped items; and 'getting lost' when switching between checklists [15]. Thus, electronic procedures and checklists have become increasingly integrated into aerospace, improving information processing for operators, and reducing their workload, errors, and response times [16].

1.3   Validation Study – Level of Integration of Procedures and System Displays

Effective design of electronic procedures supports the activities that help maintain and improve operator SA: situation management, control, and understanding of the situation [17]. Typically, operators set and verify parameters by scrutinizing one or more graphical displays presenting information on system state (e.g., items off/on; valves open/closed; landing gear lowered/raised). However, when there is limited space on existing displays on which to present graphical information of system state, or when additional displays are not available, operators may need to explicitly present desired information on an existing display (with the attendant potential for display clutter), or they may otherwise need to search for it. A key principle of display design involves the interface between the display of more than one information source and the integration or understanding of that information [18]. Garner [19] reported on what later became the theoretical foundation for the *compatibility of proximity* principle. To the extent that multiple channels of information must be *mentally* integrated or proximate, the principle asserts that the information should also be *physically* integrated or proximate. However, it is unclear whether operators working with largely textual information additionally require a graphical component to represent system state, at least in terms of maintaining sufficient awareness of actions previously made, actions currently being made, and/or actions yet to be made. Thus, several questions remain in terms of operator cognitive burden and SA in the context of electronic procedures. Can operators maintain sufficient awareness of and performance with electronic procedures *without* an additional information source providing a graphical depiction of system state? Put another way, do operators need to refer to a graphical component to maintain sufficient awareness and acceptable workload, or can they complete the procedures effectively using textual information alone? Note that subjects in a previous study indicated a desire for graphical displays during the execution of system management procedures ([20]). If operators *do* require graphical representation of system state, how is performance, SA, and workload impacted when the graphics are persistent versus when they are presented via explicit operator action to display them? Procedures that are closely integrated with relevant system displays (e.g., shown adjacent to one another or intermixed) may provide necessary context and improve performance, or they may increase complexity of eye scans and draw attentional resources unnecessarily. Procedures that are loosely integrated with relevant system displays (or not accompanied by displays at all) may provide efficiencies by reducing eye scans and attention draws, or may hinder performance due to lack of context. Addressing these questions requires investigation of operator performance using electronic procedures with varying levels of information integration. The objective of the current study was to investigate how varying levels of information integration impacted performance in the context of a primary procedural task (including malfunction procedures), and a secondary vehicle ascent monitoring task. The results of this and similar studies help system designers identify information integration design solutions for future electronic procedures systems.

The hypothesis for this study was as follows:

- As the level of integration of system and procedural information increases
- SA and usability will increase, and workload will decrease.

This study, conducted in the Crew Interface Rapid Prototyping Lab (RPL) at Johnson Space Center, involved subjects learning and performing procedural tasks, including fault management, with a high-fidelity notional electronic procedures system at NASA. Because some significant system pre-training was required, the study was conducted in two separate sessions: Session 1-Training, and Session 2-Performance. The primary objective of Session 1 was to familiarize subjects with the experimental task, including electronic procedures, spacecraft displays, and fault management (executing procedures to resolve system caution and warning conditions). Session 2 included a brief refresher training, followed by the experimental session. The time between Session 1 and Session 2 was an average of 1.95 days (range 0–8 days).

2 Method

Twenty astronaut-like subjects (13 male, 7 female) participated in Mission Control scenario-based simulations using electronic procedures. Subjects were recruited through the NASA Johnson Space Center (JSC) Human Test Subject Facility, and were crew-like in age, education, vision, hearing, and fitness level. The NASA Johnson

Space Center Crew Interface Rapid Prototyping Laboratory (RPL) developed a notional spacecraft procedures system, and a mission control/fault management simulation for use in this study. For the purpose of the study, the system was called *Zeta Mission Control*. The study was conducted in the RPL test facility, using a spacecraft workstation mockup (see Fig. 2). The workstation provided three display units, but we only used the leftmost two for this study. Each portrait-oriented display unit consisted of two display areas (upper and lower). Systems displays and procedures were viewable on either the top or bottom of each display unit.

Fig. 2. Experimental setup for the Zeta Mission Control scenario

As a Zeta Mission Controller, each subject was responsible for performing procedural tasks, responding to malfunctions and working malfunction procedures, and monitoring shuttle launches. Figure 3 shows an example set of experimental procedures and displays. Subjects performed procedural tasks in three different experimental configurations/conditions:

1. Procedures-only: In this condition, subjects were provided procedural instructions with telemetry values alongside the procedure step in the upper display region. There were no graphical system displays shown in this configuration.
2. Serial Procedures: In this condition, subjects saw telemetry values alongside the procedure steps, AND could see the value within the context of a graphical system display – but only when sending commands or on request. This configuration used only the upper display region to show subjects the procedures OR the graphical system displays (serially – one at a time).
3. Simultaneous Procedures: In this condition, subjects saw telemetry values alongside the procedural steps in the upper display region, AND saw the values within the context of a graphical system display shown in the lower display region (simultaneously – both at the same time).

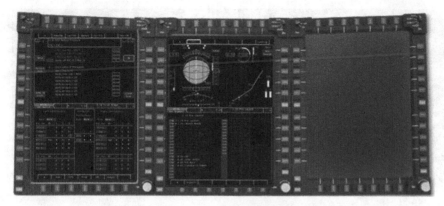

Fig. 3. Example display unit (left) showing an electronic procedure on the top and the associated system display on the bottom. Primary flight display (middle) for monitoring Zeta vehicle launches (secondary task). Far right display unit was not used for this study

All procedure/display conditions were shown on the leftmost display unit, and the primary flight display was always shown on the top of the display unit that was adjacent. Lower display areas that were not part of a tested configuration always showed a system display that was irrelevant to the task. This study used a 3×1 within-subjects design. There were three levels of procedure/display integration: *procedures-only, serial procedures,* and *simultaneous procedures.* The *procedures-only* condition showed just the procedures with telemetry values on the top half of the display unit, with an unrelated display on the bottom half. The *serial procedures* condition presented procedures with telemetry values, and the associated system display one at a time (toggle fashion on the top half of the display unit), with an unrelated display on the bottom half. The *simultaneous procedures* condition presented procedures on the top-half of the display unit and the associated graphical system display on the bottom-half of the display unit. Several human performance measures were collected for each block of performance: accuracy, SA, and cognitive workload. We measured accuracy of completing procedures by documenting errors committed during procedure execution. These include incorrect selections in branching, and skipped steps. Status Checks from the Mission Control Center (MCC) were actually SA queries (modified Situation Presence Assessment Method-SPAM; [21]). Subjects saw Status Check steps embedded in their electronic procedures (see Fig. 4), and were instructed to respond to queries on an MCC computer. To minimize the ability of subjects to predict when queries were coming, two-thirds of the Status Check instructions led to queries, and one-third of the Status Check instructions led to an alternate task (i.e., complete the Bedford workload scale).

Queries to assess SA were developed for each procedure. Most queries related to the systems being configured in the procedures, while some queries asked about the secondary task (i.e., how many vehicles had launched). The goal in query construction was to attempt to cover Level 1 SA (perceiving relevant elements in the environment), Level 2 SA (understanding their meaning), and Level 3 SA (projecting their status into the near future) [22, 23], and to make questions as equivalent as possible in terms of

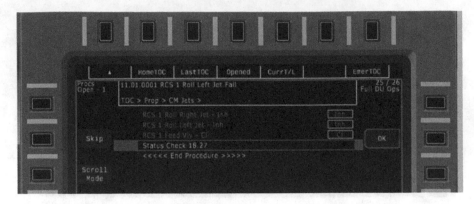

Fig. 4. Example of Status Check embedded into procedure

difficulty and time since the event in question occurred. All query questions were presented as Yes/No or multiple choice questions. In addition to the SPAM measures of SA, subjects were asked to rate their perceived situation awareness after each condition using a 4-point scale. The question provided a definition of SA, and then asked for a 1–4 rating for the question: What was your situation awareness with respect to the system? Alongside determining differences in SA due to level of procedure/display integration, a secondary question of interest was: how often are subjects actually looking at the associated system display in the Simultaneous Procedures condition? They do not actually need to view the display to perform the task because the relevant data are displayed in the telemetry boxes within the procedure; but, how much are the system displays actually being used? A Tobii mobile eye tracking system was used to make a high-level assessment. The system consisted of Tobii Pro Glasses 2 eye tracking eye glasses worn by the subject throughout the experimental session. Due to availability constraints with the eye tracker, only 10 of the 20 subjects wore the eye glasses. For eye tracking data, areas-of-interest (AOIs) were defined within the Tobii Pro Lab Analyzer edition software in order to analyze the observed visual attention being given to each area. Four AOIs were defined: the top-left, the top-right, the bottom-left, and the bottom-right display. Procedures or displays were shown within these areas of interest, depending on condition. Subjective workload was measured at the end of each condition using the Bedford Workload Scale (note: these were in addition to Bedford scales administered during the conditions as "alternate" Status Check tasks). This scale was selected because it is quick to administer and is the requirement of record for workload on the Orion and commercial spacecraft programs at NASA. During the training session, subjects were asked to complete a consent form and view a short presentation describing the study and the general tasks they would be asked to perform. They were introduced to the Zeta Mission Control Workstation and associated controls. They were shown the procedures and displays they would be using, as well as the three different procedure/display configurations with which they would be performing: Procedures-only, Serial Procedures, and Simultaneous Procedures. A malfunction alarm tone and a launch alarm tone were demonstrated, as well as the appropriate responses to the tones. Subjects were trained to silence the tone, and then either work the malfunction procedure associated with the alarm (if an alarm tone), or monitor the primary flight display if a launch alarm (see Fig. 4). Subjects were trained to confirm that the Zeta shuttle left the launch pad, g-levels remained "nominal" (green), and the vehicle stayed on the proper trajectory during ascent (triangle following trajectory line). The simulation was configured such that these cases were always true, but the subject did not know this. Subjects were also presented with definitions of SA and cognitive workload, and trained on how each would be assessed. They were also told that in addition to working through procedures and answering MCC Status Checks, one of their tasks was to monitor Zeta vehicle launches using the primary flight display, which was always located on the second display unit from the left.

During training, subjects were shown the fNIRS they would be wearing (Charles River Analytics' fNIRS Pioneer™), and the eye tracking glasses, if applicable (10 of the 20 subjects wore eye tracking glasses). Subjects completed a training procedure set with each display configuration. During training sets, subjects were encouraged to "think aloud" as they progressed through the procedures, and were free to ask any questions they had. Feedback from the test conductor was provided for the purpose of enhanced training. Subjects were asked to perform tasks as quickly and as accurately as possible. In the experimental session, each subject was first prepped for the study (i.e., eye tracking glasses donned and the fNIRS Pioneer sensor placed on the head with a headband). The subject was told the procedure/display condition they would be working with, and then began working through procedures for a nominal Guidance/ Navigation/Control task. Periodically, at points predetermined in the scenario by the investigators, an alarm tone sounded, indicating a system fault or launch of a vehicle. As trained, the subject silenced the tone, and either started performing the relevant malfunction procedure or checked the flight display to confirm the launch, fuel, and path. As the subject reached a Status Check instruction, they turned to the MCC computer and completed the SA queries. Alarm tones did not overlap, and the number of launch alarms and malfunctions was the same across conditions. On completion of the associated malfunction procedures, subjects returned to and completed the nominal task that they had been working on prior to the fault. Launch tones also occurred throughout the session, but never overlapped with malfunction alarm tones. Procedure/display configuration conditions were counterbalanced, and the experimental session lasted approximately 1.5 h. After each condition, subjects completed a questionnaire and took a brief break. Once all conditions were completed, an end-of-session questionnaire was administered to collect overall thoughts on the three Procedure/Display configurations.

2.1   CAPT PICARD Workload Measurement

After arriving for the experimental session, subjects were outfitted with the fNIRS Pioneer sensor and eye tracker glasses (if applicable). They performed a brief baseline data collection session to calibrate CAPT PICARD, which involved performing an n-back working memory task [24]. In this type of task, increases in difficulty show a corresponding change in the neurophysiology of the subject. To perform the n-back task, the subject watches a sequence of letters displayed on the screen, and they must indicate with a button press if the current letter matches the one presented some number back. For the easy condition, they respond "yes" (left mouse button) when the current letter matches the letter seen one before (1-back), else they respond "no" (right mouse button). For the next level of difficulty they respond "yes" when the current letter matches the letter seen two before (2-back), and so on. For each increase, the subject must hold more information in working memory to perform well on the task. Subjects in the study completed a 1-back task and a 3-back task as part of the CAPT PICARD baseline.

3   Results

3.1   Spam SA Assessments

Procedural errors in branching occurred to a small degree during the training session, but were not observed during experimental trials, and no steps were skipped. Mean response times and accuracy levels are presented in Table 1. Using repeated-measures analysis of variance (ANOVAs) on log-transferred response time, we show that SA did not significantly differ across the three procedure/display conditions in terms of average time to submit a response to SA queries using modified SPAM, $F(2, 38) = 0.24$, $p = 0.79$. Accuracy rates were not analyzed because they were similar and high.

Table 1. Mean response time and accuracy for queries by condition

Condition	Mean response time in seconds (SD)	Mean percent correct
Procedures only	14.1 (5.5)	91.3
Serial procedures	13.8 (5.7)	90.0
Simultaneous procedures	14.9 (8.4)	89.6

3.2   Spam SA Assessments

Subjective self-ratings of SA are shown in Table 2. We found a significant difference in self-rated SA using a repeated measures ANOVA, $(F(2, 56) = 7.5, p < 0.01)$. Follow-up pairwise comparisons indicate significant differences between Serial and Simultaneous conditions $(t = 3.56, p < 0.003)$, and Procedures-only and Simultaneous conditions $(t = 4.87, p = 0.0001)$, but not Serial and Procedures-only conditions $(t = 2.18, p = 0.042)$ at a Bonferroni-corrected p-value threshold of 0.016. So, while the SPAM method did not reveal a significant difference in SA, a self-rating of SA did show the highest subjectively rated SA was found in the Simultaneous Procedures condition. This supports the importance of an always-visible graphical display during a procedural task.

Table 2. Self-rated SA

Condition	Self-rated SA
Procedures only	1.9
Serial procedures	2.3
Simultaneous procedures	2.9

Questions from the System Usability Scale (SUS) were included as part of the end-of-block questionnaire; thus, subjects rated each Procedure/Display configuration for usability after interacting with the configuration. Figure 5 shows the average SUS scores by condition. Simultaneous Procedures had the highest usability score, followed by the Procedures-only condition. The Serial condition had the lowest usability score, most likely due to the fact that the configuration automatically jumped to the system display for commands, often leaving the subject disoriented. The lack of user control of this function is a usability issue, and there were numerous comments from subjects to that effect.

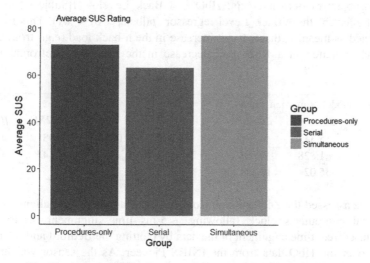

Fig. 5. Average SUS score across Procedure/Display conditions

At the end of the study session, subjects were asked for their preference regarding the different configurations with which they interacted during the study. Three participants preferred the Procedures only condition, zero preferred the Serial condition, and 17 preferred the Simultaneous Procedures condition. Participants preferred the Simultaneous Procedures condition significantly more often than either the Serial Procedures or the Procedures-only conditions $(\chi^2 = 24.7, p < 0.0001$; post-hoc $p's < 0.05)$ (Table 3).

Table 3. Self-rated preference

Condition	Self-rated preference
Procedures only	3
Serial procedures	0
Simultaneous procedures	17

The Bedford questionnaire was administered at the end of each condition. The workload ratings were identical across conditions (3.55); therefore, they were not analyzed further.

3.3 Neurophysiology: CAPT PICARD Results

First, Charles River assessed data from the collected baseline n-back task to ensure that it showed the typical trend of progressively increasing HbO as the n-back level increases. Indeed, we found the typical trend wherein HbO changed as a function of n-back level. A linear mixed-effects model analysis (linear effect of n-back level on HbO with a by-subject random intercept: HbO ~ NBack_Level + (1|Subject)) revealed a significant effect of the NBack_Level regressor ($p(|beta| > 0) < .01$). This result may be interpreted as meaning that a 1 unit increase in the n-back load (e.g., from a 2-back to a 3-back) resulted in a 1.828 unit decrease in the measured prefrontal HbO on average.

Mixed linear model regression results						
	Coef.	*Std. Err.*	*z*	*P > |z|*	*[0.025]*	*[0.975]*
Intercept	15.343	4.726	3.247	0.001	6.08	24.606
Load	−1.828	0.008	−240.187	0	−1.843	−1.813
Subject RE	335.02	18.311				

Next, we assessed the Zeta Mission Control task. Data were time-aligned using the experimental computer's clock (allowing ∼5 ms time alignment accuracy). We assessed the correct time alignment of the data by plotting the Bedford and SA question responses over the HbO data from the fNIRS Pioneer. As the sensor was turned off between blocks, if data is time aligned correctly, Bedford and SA question responses should only occur during periods where the sensor was turned on and data were received. Figure 6 shows that this is indeed the case, verifying broadly that time

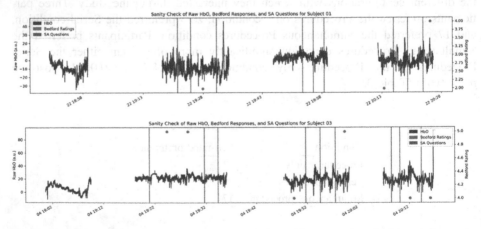

Fig. 6. Example graphs showing time-aligned data for subject 1 (top) and subject 3 (bottom). Y axis provides time of data in GMT. First block of data (blue trace at beginnings of recordings without Bedford or SA questions) was during the baseline N-back data (not analyzed here). (Color figure online)

alignment was performed adequately. Data from subject 1 is shown on the top and from subject 3 is shown on the bottom.

Figure 7 shows a comparison of the Bedford scores collected during the Status Check portions of each condition (as opposed to the end-of-condition Bedford scores), the SA questions, and the fNIRS data. Bedford scores elucidated during the task were numerically lower (least cognitive workload) for the Procedures-only condition than for the other conditions; however, this finding is not significant (single tailed t-test) and resides in the tails of the distributions (mean 3.28 for Procedures-only vs 3.47 and 3.41, out of 10). Bedford scores given immediately following each condition do not show a difference between conditions. This story is mirrored by the SA questions, where the Procedures-only had (again, non-significant) slightly higher accuracy for SA questions (91.2% vs 90.0% and 90.0%). Physiological variables output by CAPT PICARD show a similar trend. We computed the mean HbO, deoxygenated blood volume (HbR), and heart rate for each block (consisting of a user using one conditions' interface style once). The Procedures-only condition resulted in a numerically lower HbO and numerically higher HbR value (indicating lower prefrontal blood oxygenation) for the Procedures-only condition relative to the other conditions.

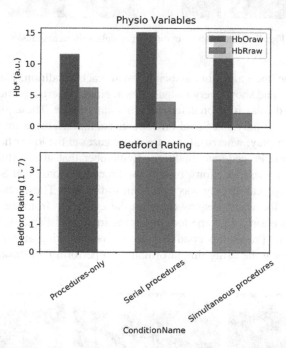

Fig. 7. Comparison of physiological and behavioral variables across conditions. Physiological variables show a significant difference between the Procedures-only condition and the other conditions, in accordance with the (non-significant) behaviorally-measured increased workload in the Procedures-only condition. Higher HbO indicates higher prefrontal blood oxygenation, which suggests greater workload.

Eye Tracking Results

A number of pre-calculated eye tracking metrics were extracted and analyzed. The primary metric of interest was "visits", or the period of time a participant spends looking at a pre-defined AOI. Figure 8 shows heat maps for each condition. The heat maps are created by averaging the amount of overlapping viewing activity on areas of the screen, with some activity being represented by a green color, moderate activity by a yellow color, and a high amount of activity by a red color. The two concentrated red-yellow areas on the upper-leftmost display unit in each condition represent frequent eye fixations on the procedural instruction, and the telemetry value to the right of each instruction.

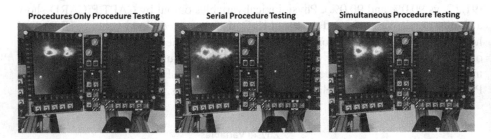

Fig. 8. Heat map of eye tracking data split across tasks

Figure 9 shows the gaze plots associated with each condition. Gaze plots are not based on averages, and show every instance (each circle) where a participant 'gazed' at an area within the display for a predetermined amount of time. These plots clearly show that in the Procedures-only and Serial Procedures conditions, there are few gazes to the lower half of the display; whereas, there are many gazes at the lower half of the display in the Simultaneous Procedures condition. Remember that all conditions provided a graphical display in these locations, but for the Procedures-only and Serial Procedures conditions, the graphical display was irrelevant to the task. These results demonstrate that it is not only the visual display drawing the eye away from the procedural task (although subjects clearly do some looking at the irrelevant display), but that subjects in the Simultaneous Procedures condition are actively looking at the lower graphical display, and appear to be using this information to perform their task.

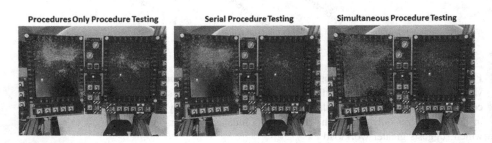

Fig. 9. Gaze Plots of eye tracking data split across tasks.

We extracted total counts, total durations, and average durations of participant visits to each of the four display AOIs from Tobii Pro Lab software, and compared these across conditions using repeated-measures ANOVA. Post-hoc Tukey tests where significant. Significant findings are reported here, grouped by AOI. Since the primary objectives of this research are concerned with the importance of a graphical display, only results associated with graphical displays are presented below.

The graphical display was shown to result in significantly greater overall attention placed upon it by the Simultaneous condition than either the Serial or Procedures-only conditions. Repeated-measures ANOVAs and post-hoc Tukey tests showed that subjects in the Simultaneous condition had more visual interaction with the graphical display than the other conditions, in the form of more total visits ($F(2, 18) = 29.73$, $p < 0.05$); higher total visit duration ($F(2, 18) = 19.08$, $p < 0.05$); and higher average visit duration ($F(2, 18) = 11.29$, $p < 0.05$). This was to be expected because, the simultaneous condition is the only condition which had a graphical display in the lower display that was relevant to task. However, what makes this finding particularly interesting is that subjects did NOT have to refer to the graphical display in order to interact with procedures or answer SA queries accurately. All the information they needed was in the form of the telemetry boxes to the right of each instruction. The graphical display mirrored the alphanumeric telemetry data shown in the procedures. Yet, subjects heavily used the graphical information in their task. Subjective comments in the questionnaires confirm that the graphical information provided important context and gave subjects confidence in their understanding of the system. The Primary Flight Display was used to monitor Zeta shuttle launches and was considered a secondary task. Subjects had shorter average visits to the primary flight display while performing in the Serial Procedures condition, than while performing in the Procedures-only

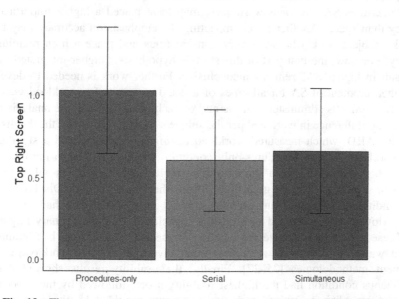

Fig. 10. The average visit duration (seconds) within the Primary Flight Display

condition, and marginally shorter average visits than while performing in the Simultaneous Procedures condition ($F(2, 18) = 5.23$, $p < 0.05$). These differences are illustrated in Fig. 10. While this study did not show a difference in workload using the Bedford Workload Scale, these eye tracking results may be an indication that the Serial Procedures condition resulted in the highest workload (i.e., there was more to look at and a more difficult condition), followed by the Simultaneous Procedures (i.e., there was more to look at and a less difficult-no toggling and no memory required). The Procedures-only condition may have resulted in the lowest workload (i.e., only text procedures to look at), leaving greater spare capacity for visually checking the Primary Flight Display. The CAPT PICARD workload data support this conclusion as well, finding the lowest blood oxygenation for the Procedure-only condition indicating a decrease in cognitive workload.

4 Conclusion and Discussion

The purpose of this study was to investigate how varying levels of information integration impacted performance in the context of a primary procedural task (including malfunction procedures), and a secondary vehicle ascent monitoring task. It was hypothesized that as the level of integration of system and procedural information increases SA will increase, and cognitive workload and usability will improve. The study was also an opportunity to validate Charles River's CAPT PICARD workload tool. The results for SA compared across the three conditions, Simultaneous (high integration), Serial (moderate integration), and Procedures-only (no integration), show mixed results. The results for SPAM showed no significant difference between the three conditions; however, self-rated SA by the subjects indicated that they perceived the highest SA in the Simultaneous condition. The mixed results could be due to how SPAM measures SA, and/or how subjects may have placed a higher importance on accuracy than speed. Another factor impacting the emphasis on accuracy may be the crew-like subject pool, who were very conscientious and place a high premium on accuracy. For now, the first part of this study's hypothesis, "higher integrated system will result in higher SA" remains inconclusive. Further work is needed to develop a robust measurement of SA for all types of tasks. The results for workload were also mixed. Bedford was administered at the end of each condition, and the analysis found there was no difference in workload per the subjects' ratings. However, the results from CAPT PICARD, which measured workload *during* the task, showed a significantly lower workload for the Procedures-only condition when compared to the other conditions. The eye tracking data show similar results. The Procedures-only condition resulted in significantly more gaze time toward the Primary Flight Display than the other conditions. This may suggest that subjects in this condition had more spare capacity (lower workload), and so used that capacity to view the Primary Flight Display. These results demonstrate the benefits of measuring workload in a number of different ways, as well as the need for further investigations into workload and SA assessment methodologies [25–27]. Finally, the usability results showed that the Simultaneous condition had the highest usability score, followed by the Procedures-only condition. This is in line with the comments provided by the subjects. It is interesting that the Serial condition, which also provided a graphical display, scored lower than the Procedures-only condition. Reviewing the subjects' comments, it appears that the Serial condition's approach of automatically jumping to the graphical display when a procedure step was a command was an issue, and was disorienting. Subjects no longer felt in control of the system. Based on self-rated SA, we can partially confirm the hypothesis. There is indication that the Simultaneous condition (high integration) provided subjects with a higher sense of SA, and similarly, usability was highest with the Simultaneous condition. However, real-time assessment of workload using CAPT PICARD along with eye tracking data did not show workload being the lowest in the simultaneous condition. This could be due to there being more information to process by the participants (e.g., telemetry data, graphical representation of the procedure and the primary flight display showing the launching of vehicles). Given that we found the expected correlation in the neurophysiological data and the ground-truth cognitive workload assessment task (the n-back), the sensor data for astronaut-like test subjects and data collected in an operational lab at NASA are consistent with that in the existing literature (i.e., change in HbO correlated with n-back difficulty level; [28]). However, we plan to continue to test this system under new conditions and new populations in future studies. There are a few limitations to the present study. The SA questions used in the present study did not include an even distribution from each of the 3 SA levels [19], making comparison between levels difficult. Additionally, the electronic procedures task used in the study may not have provided a level of variability in workload that would be best suited for validation. Finally, a larger sample size would provide more robust results for each of the measures investigated. The value of workload assessment during developmental testing of a new system is that it allows for early identification of features and designs that result in high workload. When issues are identified early, redesigns are more feasible and less costly. Real-time workload data could provide the inputs needed to drive future adaptive displays (e.g., if an astronaut is experiencing high cognitive workload, this data could cue an option for more simplified displays). Results from this study represent a step towards validation of a flexible fNIRS-based workload tool. A future vision for real-time, unobtrusive measurement of workload is also to use it in an operational spaceflight environment. Therefore, it is critically important to continue to test and mature tools for unobtrusive measures of human performance like CAPT PICARD.

Acknowledgements. This work was funded by the NASA Small Business Innovation and Research funding (contract numbers NNX15CJ17P and NNX16CJ08C) and the NASA Human Research Program (contract number NNJ15HK11B). The study, under the direction of Kritina Holden, was performed at the NASA Johnson Space Center. The authors would like to thank Debra Schreckenghost, Christopher Hamblin, Lee Morin, and Camille Peres for their technical contributions. A special thanks to Patrick Laport, who developed the procedures and simulation software for the study. Finally, the authors would like to thank Lee Morin for use of the Crew Interface Rapid Prototyping Laboratory at the NASA Johnson Space Center.

References

1. Kerick, S.E., Oie, K.S., McDowell, K.: Assessment of EEG signal quality in motion environments. Army Research Lab Aberdeen Proving Ground MD Human Research and Engineering Directorate (2009)
2. Boas, D.A., Elwell, C.E., Ferrari, M., Taga, G.: Twenty years of functional near-infrared spectroscopy: introduction for the special issue. NeuroImage **85**, 1–5 (2014). https://doi.org/10.1016/j.neuroimage.2013.11.033
3. Ferrari, M., Quaresima, V.: A brief review on the history of human functional near-infrared spectroscopy (fNIRS) development and fields of application. NeuroImage **63**(2), 921–935 (2012). https://doi.org/10.1016/j.neuroimage.2012.03.049
4. Ayaz, H., Shewokis, P.A., Bunce, S., Izzetoglu, K., Willems, B., Onaral, B.: Optical brain monitoring for operator training and mental workload assessment. NeuroImage **59**(1), 36–47 (2012). https://doi.org/10.1016/j.neuroimage.2011.06.023
5. Bunce, S.C., et al.: Implementation of fNIRS for monitoring levels of expertise and mental workload. In: Schmorrow, D.D., Fidopiastis, C.M. (eds.) Foundations of Augmented Cognition. Directing the Future of Adaptive Systems, vol. 6780, pp. 13–22. Springer, Heidelberg (2011). https://doi.org/10.1007/978-3-642-21852-1_2
6. Bracken, B.K., Elkin-Frankston, S., Palmon, N., Farry, M., de B Frederick, B.: A system to monitor cognitive workload in naturalistic high-motion environments (2017)
7. Bracken, B.K., Palmon, N., Elkin-Frankston, S., Silva, F.: Portable, Durable, Rugged, Functional Near-Infrared Spectroscopy (fNIRS) Sensor, February 2018
8. Molavi, B., Dumont, G.A.: Wavelet-based motion artifact removal for functional near-infrared spectroscopy. Physiol. Meas. **33**(2), 259 (2012)
9. Metz, A.J., Wolf, M., Achermann, P., Scholkmann, F.: A new approach for automatic removal of movement artifacts in near-infrared spectroscopy time series by means of acceleration data. Algorithms **8**(4), 1052–1075 (2015)
10. Scholkmann, F., Spichtig, S., Muehlemann, T., Wolf, M.: How to detect and reduce movement artifacts in near-infrared imaging using moving standard deviation and spline interpolation. Physiol. Meas. **31**(5), 649–662 (2010). https://doi.org/10.1088/0967-3334/31/5/004
11. Cao, A., Chintamani, K.K., Pandya, A.K., Ellis, R.D.: NASA TLX: software for assessing subjective mental workload. Behav. Res. Methods **41**(1), 113–117 (2009)
12. Murphy, K.P.: Dynamic Bayesian Networks: Representation, Inference and Learning (Dissertation). University of California, Berkeley (2002). https://ibug.doc.ic.ac.uk/media/uploads/documents/courses/DBN-PhDthesis-LongTutorail-Murphy.pdf
13. Cox, Z., Pfautz, J.: Causal influence models: a method for simplifying construction of bayesian networks. Charles River Analytics Inc. (2007)
14. Pélegrin, C.: The never-ending story of proceduralization in aviation. In: Trapping Safety into Rules, pp. 31–44. CRC Press, Boca Raton (2017)
15. Palmer, E., Degani, A.: Electronic checklists: evaluation of two levels of automation. In: Proceedings of the Sixth Symposium on Aviation Psychology, pp. 178–183. The Ohio State University Columbus (1991)
16. Myers III, P.L.: Commercial aircraft electronic checklists: benefits and challenges (literature review). Int. J. Aviat. Aeronaut. Aerosp. **3**(1), 1 (2016)
17. de Brito, G., Boy, G.: Situation awareness and procedure following. In: CSAPC, vol. 99, pp. 9–14 (1999)
18. Barnett, B.J., Wickens, C.D.: Display proximity in multicue information integration: the benefits of boxes. Hum. Factors **30**(1), 15–24 (1988)

19. Garner, W.R.: The stimulus in information processing. In: Moskowitz, H.R., Scharf, B., Stevens, J.C. (eds.) Sensation and Measurement, pp. 77–90. Springer, Dordrecht (1974). https://doi.org/10.1007/978-94-010-2245-3_7

20. Schreckenghost, D., Milam, T., Billman, D.: Human performance with procedure automation to manage spacecraft systems. In: 2014 IEEE Aerospace Conference, pp. 1–16. IEEE (2014)

21. Durso, F.T., Dattel, A.R., Banbury, S., Tremblay, S.: SPAM: the real-time assessment of SA. A Cogn. Approach Situat. Aware. Theory Appl. 1, 137–154 (2004)

22. Endsley, M.R.: Measurement of situation awareness in dynamic systems. Hum. Factors 37(1), 65–84 (1995)

23. Ericsson, K.A., Hoffman, R.R., Kozbelt, A., Williams, A.M.: The Cambridge Handbook of Expertise and Expert Performance. Cambridge University Press, Cambridge (2018)

24. Kirchner, W.K.: Age differences in short-term retention of rapidly changing information. J. Exp. Psychol. 55(4), 352 (1958)

25. Moustafa, K., Luz, S., Longo, L.: Assessment of mental workload: a comparison of machine learning methods and subjective assessment techniques. In: Longo, L., Leva, M. (eds.) H-WORKLOAD 2017. CCIS, vol. 726, pp. 30–50. Springer, Cham (2017). https://doi.org/10.1007/978-3-319-61061-0_3

26. Vidulich, M.A., Tsang, P.S.: The confluence of situation awareness and mental workload for adaptable human–machine systems. J. Cogn. Eng. Decis. Making 9(1), 95–97 (2015)

27. Muñoz-de-Escalona, E., Cañas, J.J.: Latency differences between mental workload measures in detecting workload changes. In: Longo, L., Leva, M. (eds.) H-WORKLOAD 2018. CCIS, vol. 1012, pp. 131–146. Springer, Cham (2018). https://doi.org/10.1007/978-3-030-14273-5_8

28. Herff, C., Heger, D., Fortmann, O., Hennrich, J., Putze, F., Schultz, T.: Mental workload during n-back task—quantified in the prefrontal cortex using fNIRS. Front. Hum. Neurosci. 7 (2014). http://doi.org/10.3389/fnhum.2013.00935

COMETA: An Air Traffic Controller's Mental Workload Model for Calculating and Predicting Demand and Capacity Balancing

Patricia López de Frutos[1]([⊠]), Rubén Rodríguez Rodríguez[1],
Danlin Zheng Zhang[1], Shutao Zheng[1], José Juan Cañas[2],
and Enrique Muñoz-de-Escalona[2]

[1] CRIDA A.I.E. ATM R&D + Innovation Reference Centre, Madrid, Spain
{pmldefrutos, rrodriguezr, dzheng,
szheng}@e-crida.enaire.es
[2] Mind, Brain and Behaviour Research Centre, University of Granada,
Granada, Spain
{delagado, enriquemef}@ugr.es

Abstract. In ATM (Air Traffic Management), traffic and environment are not important by themselves. The most important factor is the cognitive work performed by the air traffic controller (ATCo). As detailed mental pictures can overcome ATCo's limited attentional resources (causing Mental Overload), she/he can use internal strategies called abstractions to mitigate the cognitive complexity of the control task. This paper gathers the modelling, automation and preliminary calibration of the Cognitive Complexity concept. The primary purpose of this model is to support the ATM planning roles to detect imbalances and make decisions regarding the best DCB (Demand and Capacity Balancing) measures to resolve hotspots. The four parameters selected that provide meaningful operational information to mitigate cognitive complexity are Standard Flow Interactions, Flights out of Standard Flows, Potential Crossings and Flights in Evolution. The model has been integrated into a DCB prototype within the SESAR (Single European Sky ATM Research) 2020 Wave 1 during Real Time Simulations.

Keywords: Cognitive complexity · Workload model · Air Traffic Controller · Demand and Capacity Balancing

1 Introduction

Air Traffic Controller (ATCo) Mental Workload (MWL) is likely to remain the single greatest functional limitation on the capacity of the ATM (Air Traffic Management) System [1]. The consensus view among the ATM research and operational communities is that complexity drives ATCo MWL [2]. This is one of the main reasons to use complexity data as additional information to support the ATM planning roles in order to detect imbalances and make decisions regarding the best DCB (Demand and

© Springer Nature Switzerland AG 2019
L. Longo and M. C. Leva (Eds.): H-WORKLOAD 2019, CCIS 1107, pp. 85–104, 2019.
https://doi.org/10.1007/978-3-030-32423-0_6

Capacity Balancing) measures to resolve hotspots. Complexity features are a relevant topic of scientific research and theory in many academic and applied fields [3]. However, a review of this research in the ATC domain shows different approaches to the complexity issue and its assessment [4]:

- *Algorithmic Approach:* It is assumed that the complexity of the task can be calculated directly from the parameters of the environment [5]. Over the years, some formulas have calculated complexity based on parameters such as occupancy (number of aircraft in the sector) or traffic density. In this approach, the ATCo is not considered in the definition of complexity. Complexity is defined only by the traffic and the environmental conditions where the task is performed.
- *Behavioral (Activity) Approach:* In this case, complexity is defined and measured from the observable behavior of the ATCo without any reference to the cognitive processing of traffic and environmental parameters. While it is assumed that ATCo behavior is the result of the cognitive processing of traffic and environment parameters, no attempt is made to model his/her cognitive processing. For example, the authors of [6] proposed a method of measuring complexity from the actions of the ATCo and the authors of [7] defined another method where complexity is calculated from the commands issued by the ATCo.
- *Cognitive System Approach* [8]: In this approach, the traffic and operational parameters are not important by themselves, but rather the way in which the ATCo adjusts his/her decision-making strategies according to the parameters of traffic and environment. This approach is built around the concept of a cognitive system. This type of system performs the cognitive work of knowing, understanding, planning, deciding, problem-solving, analyzing, synthesizing, assessing, and judging as they are fully integrated with perceiving and acting. This research path uses the hypothesis that the ATCo is part of an entity that performs cognitive work, taking the parameters of traffic and operational environment into account.

The current study searches for an approach to predict complexity in the ATM planning phase (namely some hours before the operation) complying with the following requirements (Req.): (Req. 1) the ATCo MWL limitations have to be taken into consideration; (Req. 2) the complexity assessment method should be adequate for automation; (Req. 3) the assessment should provide meaningful operational information for the ATM planning roles to make decisions on which DCB measure is the most appropriate (diagnostic capability); (Req. 4) the inputs to assess complexity must be available in a real environment at the planning phase. Bearing all the requirements expressed in mind, the Algorithmic Approach is discarded, as the ATCo is not taken on board. The Behavioral Approach, although it considers the ATCo as the center of the measurement, might not be appropriate since the inputs required for its assessment, namely the behavioral parameters, are not available during the planning phase. Regarding the Cognitive System Approach, it takes the ATCo as the center of the assessment and underpins on a working mental model that can draw upon the constraints of human memory and processing limitations. In this way, the main objective of the research presented throughout this study is to build a computerized model based on a Cognitive System Approach that addresses the rest of the above-mentioned requirements: its diagnostic capability (Req. 3) and the availability of enough inputs some hours before the operation (Req. 4).

2 The COMETA Computational Model

COMETA (Cognitive ModEl for aTco workload Assessment) is a computerized model researched and developed within CRIDA (ATM R&D + Innovation Reference Centre) internal projects for ENAIRE (Spanish Air Navigation Service Provider) and for the SESAR programme, to estimate and predict the ATCo MWL. Typically, it has been used in the field of ATM complexity management concepts [9–11] and for Human Factors research [4, 12, 13]. COMETA follows a Cognitive System Approach based on an ATCo Psychological Model that considers two main components: the functional structure of the controller cognitive system and the attentional resources needed for its functioning to carry out the air traffic management tasks.

2.1 COMETA Architecture. Implementing the Structure of the ATCo Cognitive System

The cognitive system has structural components whose functions are the processing of external information, the storing of the results of that processing and the responding to the environment. The reference taken for the ATCo psychological model is proposed by [14], and it incorporates aspects to better represent the ATCo activity (Fig. 1). This model includes the levels of processing that constitute the Situation Awareness (SA) [15] (perception, comprehension, projection), Decision Making and Execution. The information received by the ATCo coming from the traffic situation and environment is processed taking into consideration the long-term memory of the controller (previous learning and experience). Then, the ATCo will project the current situation into the future to predict how the traffic will evolve and finally, he/she will make decisions to correctly perform the needed tasks.

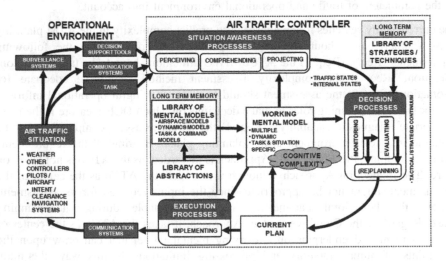

Fig. 1. Functional structure of the ATCo cognitive system [14]

2.2 COMETA Functional Architecture

The activity of the cognitive processes requires energy [16], and the performance of a task will be improved or deteriorated depending, among other factors, on the quantity and quality of the energy (attentional resources) supplied [17]. While demanded resources are those required by the task and essentially dependent on the task complexity, available resources are the ones that the ATCo has that could be used to perform the task. The available resources mainly depend on the ATCo's level of activation or alertness. The cognitive system will work with the efficiency given by the relationship between the demanded and the available resources. This relation is known as the Mental WorkLoad (MWL) [2]. The COMETA algorithm estimates the demanded resources of ATCo tasks following [18, 19] and takes into account the required cognitive processes to perform these tasks. However, the available resources calculator that is dependent on the ATCo level of activation is not yet incorporated. Figure 2 shows the COMETA functional architecture. The items inside the black box refer to what is currently developed.

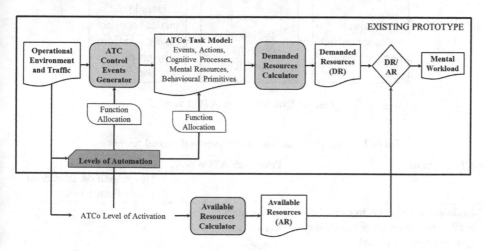

Fig. 2. COMETA functional architecture

2.3 COMETA Inputs and Outputs

COMETA needs as inputs:

- *Air Traffic Control (ATC) Events:* They are the psychological stimulus to which the ATCo responds (e.g. solve conflicts). The operational environment and the air traffic under the controller responsibility are taken into account through the ATC Events that might be acquired in real-time and post-processed or generated by means of simulations.

- *An ATCo Task Model* composed by *ATC Actions:* They are observable behaviors that can be defined as the "behavior of an actor directed to an objective" [20]. The actions are carried out with the implication of cognitive processes that consume attentional resources. In order to link the cognitive processes required to perform air traffic control actions triggered by ATC Events, Primitive ATCo Tasks [21] are defined to facilitate the psychological modelling of an ATCo interacting with the system interface. Figure 3 and Table 1 are examples of ATC Events and the link with the ATCo Task Model. In Table 1, the mental resources required by each behavioral primitive have been estimated by expert air traffic controllers.

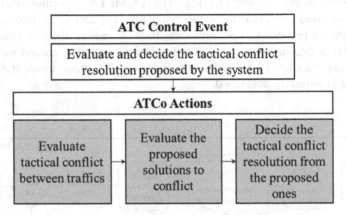

Fig. 3. Example of an ATC Event

Table 1. Example of actions, primitives and mental resources

ATCo actions	Primitives ATCo tasks	Mental resources (expressed on an interval scale from 1 to 10)
Evaluate the solutions to early conflict resolution proposed by the system (planning phase)	Fixate object	3
	Recall	4
	Recognize	5
	Select	7
	Compare	7
	Compute	8
Decide the early conflict resolution from the proposed ones (by the system)	Decide	8

As outputs, COMETA provides the Cognitive Demand of the tasks performed as a global and per cognitive process. If it is assumed that the ATCo's level of activation would be adequate to maintain an optimum performance, the available resources would be kept as fixed and the demanded resources estimation would mean MWL estimation.

2.4   Limitations of COMETA

COMETA can be considered as a human factors tool to aid in the design and analysis of complex human-machine systems. As the ATCo Task Model can be tailored according to ATCo experience, task complexity and the level of automation (function allocation), COMETA allows users to perform human factors analysis of new concepts of operation at early stages, prior to the use of human-in-the-loop experiments. Furthermore, COMETA is being currently used at CRIDA/ENAIRE to analyze the ATCo MWL using historical data. Unfortunately, this version of COMETA is not complying with Req. 3 (diagnostic capability) and Req. 4 (inputs availability) as described in Sect. 1. However, COMETA does not reflect the operational information required to understand the root causes of complexity in order to allow ATM planning actors to implement the most appropriate DCB measures. DCB measures are actions on demand (individual flight trajectories or flows) or on capacity (airspace organization) to balance both. Typical measures on demand (either for individual trajectories or flows) are Re-Routings (alternative routes or flight level modifications in order to resolve ATC capacity problems by balancing the traffic load among sectors) or CTOT allocation (Calculated Take-Off Time) measures as a delay imposed on the ground, either at the gate or on the taxiway to reduce a sector or airport overload [22]. COMETA offers the identification of mental overloads (cognitive demand) but not the complex flights or flows, so its diagnostic capability (Req. 3) for DCB measures selection is not appropriate. Regarding the inputs available (Req. 4), preliminary calibration of COMETA has been done with historical data where all inputs are easily accessible. However, in the ATM planning phase (prediction) the ATCo is not yet managing the traffic and therefore the ATC Events have to be predicted as well. The prediction of ATC Events is not an easy task and the different approaches followed in SESAR 1 did not achieve satisfactory results [9, 10]. The algorithmic needed to predict what are the most probable ATCo interventions at different operational situations implies analysis of massive empirical data and observations. Further research on the use of big data and machine learning techniques would help to solve this issue in the future. Due to these limitations, a new research area to estimate ATCo Mental Workload and Complexity is open to mitigate those abovementioned limitations. This investigation searches for an innovative tool that, taking the ATCo as the center of the system, allows the ATM planning roles to make decisions to balance demand and existing capacity.

3   Psychological Basis for the Calculation of Cognitive Complexity

In the field of ATM, Cognitive Complexity is the cognitive difficulty of controlling an air traffic situation [14]. The research carried out by [14] has been taken as the main reference for the current investigation as gathers interesting concepts for Cognitive Complexity modelling. Complexity is not directly associated to the stimuli (traffic and environment), but rather how complex it is to perceive, comprehend, project, make decisions and execute an action. In other words, the complexity is associated to the cognitive processing of the stimuli and to the interaction between the human being and the system [23]. At the center of the cognitive process model in Fig. 1, there is an ATCo working mental model, also known as a mental picture. According to [14], the working mental model supports the generation and maintenance of situation awareness as well as the multiple decision-making and implementation processes. The working mental model is a product of abstractions, mental models and other parts of their long-term memory combined with the controller's "Current Plan." It integrates the different sources of information available to the controller, including perceptual clues of the current positions of aircraft and their future intent. Working mental models can draw upon abstractions or simplified versions of a system's dynamics. Abstractions are a mean of representing the essential characteristics of a mental model in a more cogni-tively compact form that is manageable within the constraints of human memory and processing limitations. The use of abstractions provides an important mechanism for reducing the demand of the controller's attentional resources. The ATCo can mitigate and reduce cognitive complexity by changing the level of abstraction, meaning that the air traffic situation can be decomposed into simpler and easier problems. In [14], four types of simplifications of the ATCo mental picture for mitigating cognitive complexity are identified: standard flow, critical point, grouping, and responsibility. Figure 4 represents sketches on these four concepts.

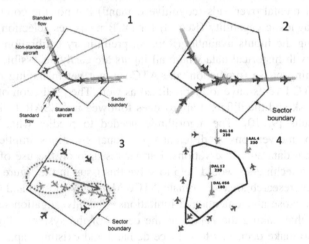

Fig. 4. Representation of the abstractions of standard flows (1), critical points (2), grouping (3) and responsibility (4) [14].

3.1 Standard Flow Abstractions

Standard flows are recurring patterns of aircraft sharing common lateral and vertical paths; in a standard flow, aircraft are typically "in-trail" of each other. Moreover, a standard flow may span multiple altitudes, include vertical behaviors such as climbs or descents, and merge and/or cross with other flows in the airspace. Standard flow abstractions simplify the working mental model used in various cognitive processes as they filter out "pre-solved" relationships between aircraft, based on the independence introduced by the arrangement of aircraft into the flow, which marks the evaluation and projection processes easier. Finally, standard flow abstractions incorporate typical commands, making identification of feasible commands and airspace available for maneuvering quicker and easier.

3.2 Critical Point Abstractions

Critical point abstractions are generalizations of high priority regions of a sector. Typically, these high priority regions, or critical points, are locations where controllers are expecting potential conflicts or other sources of recurring problems (e.g. overshooting a turn in an airway). Critical point abstractions transform multi-aircraft and multi-time step projections into a simpler projection of the time of arrival at the fixed location of the critical point. Evaluation and planning are also simplified. Critical point abstractions also support a more focused monitoring and perceiving processes.

3.3 Grouping Abstractions

This is distinct from the standard flow abstractions in the sense that the latter reflects common spatial trajectories, whilst the grouping abstraction is capturing the relative proximity of a set of actual aircraft. Generalizations of aircraft performances also provide basis for a subset of grouping abstractions, identified as performance abstractions. These include climb/descent rates, speeds, navigation capabilities, and/or willingness to penetrate turbulence.

3.4 Responsibility Abstractions

Responsibility abstractions internalize the structure's effect on the task and the delegation of portions of the task to other agents or parts of the system. Responsibility abstractions limit the scope of monitoring, evaluating and projecting processes. Airspace boundaries, eliminating aircraft from the task, are a simple example of a basis for a responsibility abstraction.

4 Parameters Selection Methodology and Formulas for Modelling ATCo's Mental Workload

The selection of parameters to model the cognitive complexity and build the current version of COMETA has followed the next three steps: (1) a literature review has been done to identify the main factors in the ATC domain influencing complexity; (2) as hypothesis, those factors that are not facilitators of ATCo abstractions have been established as sources of cognitive complexity; and (3) with ENAIRE's operational support, a set of factors has been prioritized to support ATM planning roles to make decisions regarding the application of DCB measures.

Step 1: A classification of factors influencing complexity has been proposed differentiating between those related to static features of a sector and those related to the dynamic situation of the air traffic [24]. Reference [14] takes this classification and adds a new group related to the operational restrictions (Table 2).

Table 2. Key factors reported by ATCo influencing complexity

Category	Factor
Airspace factors	Sector dimensions
	Standardized procedures
	Number and position of standard ingress/egress points
	Spatial distribution of airways
	Standard flows (number of, orientation relative to sector shape, trajectory complexity, lack of)
	Interactions between standard flows (crossing points, merge points)
	Coordination with other controllers (hand-offs, point-outs)
Traffic factors	Density of traffic (clustering, sector-wide)
	Aircraft encounters (number of, distance between aircraft, relative speed between aircraft, location of point)
	Ranges of aircraft performance
	Number of aircraft in transition (altitude/heading/speed)
	Sector transit time
	Relationship of aircraft to standard flows (presence of non-standard aircraft)
Operational constraints	Restrictions on available airspace
	Buffering capacity
	Procedural restrictions
	Communication limitations
	Wind effects (direction, strength, changes)

Step 2: Factors from Table 2 have been classified according to items that do not facilitate the four types of ATCo mental abstractions:

- Standard Flows Abstractions: those factors that would create cognitive complexity and are not facilitators for standard flows abstractions are a) the aircraft out of standard flows, and b) flights that belonging to standard flows interact between them. The flows interaction creates more complexity when more flights are involved within the flows. The type of standard flows interaction is also a source of more or less complexity. For instance, the interaction between two established flows (no evolution) is less complex than between two flows in evolution.
- Critical Points Abstractions: potential conflicts between flights in areas that are not included within a critical point category create more complexity, as they are less familiar for the involved actors. In turn, the geometric features of the potential conflict are sources of complexity for the ATCo. *Potential aircraft encounters* in areas not considered as critical create more complexity.
- Grouping Abstractions: those aircraft that cannot be grouped by a feature, meaning that they behave differently with respect to the "average", are considered as sources of cognitive complexity. For instance, *Aircraft not following Standard Procedures,* or *Not entering/exiting through standard points, Aircraft with no normal Performance Ranges, Aircraft in evolution in upper sectors (en-route),* or *Aircraft whose Transition Time in the sector is higher than the mean (slow flights).*

- Responsibility Abstractions: the more aircraft under ATCo responsibility, the higher the risk of cognitive complexity. Consequently, the *Sector Dimensions* or *the Traffic Density* may be potential parameters to take into consideration.

Transversally, there are also other factors that create cognitive complexity such as *Number of Coordinations, Communications Limitations* and *Wind Effects*.

Step 3: The final purpose of the computerized model in this study is to predict Cognitive Complexity and to understand the main reasons, given the foreseen complexity, to take actions on individual flights or flows. For this reason, all the factors influencing complexity that are referred to individual flights or flows are on top of the priority list. After a detailed analysis with ENAIRE's operational staff, the following parameters were selected: *Standard Flows Interaction, Flights out of Standard Flows, Flights in Evolution and Potential aircraft encounters, renamed as Potential Crossings*. These parameters are foreseen to be used for En-Route Sectors. These four parameters will be the contributing factors to the Cognitive Complexity computerized model.

4.1   Standard Flows Interaction. Rationale for the Formula

Cognitive Complexity associated to the standard flows interaction depends on (a) the type of flows interaction and (b) the flights involved in the interaction.

A standard flow is a set of trajectories that all together comply with common spatial characteristics during a period of time. The development process to detect air traffic flows are: (1) application of clustering algorithms to group air traffic trajectories based on common characteristics; (2) definition of flows based on the mean trajectory of grouped trajectories; (3) execution of a flow comparison algorithm to trace the progression of the same flows throughout a large period of time; (4) pattern detection to identify and predict how flows evolve over time. The calculation of flows is performed within an internal CRIDA project. Currently, the flows are calculated for all En- Route sectors of the Spanish airspace, within cycles of 28 days (AIRAC) for the entire 2017 and 2018 dataset. An example of flows in a Madrid ACC sector is shown on Fig. 5. The flows are characterized by the number of flights in the flow during the analysis period (AIRAC cycle), flow attitude (cruise, climb, descent), mean discretized trajectory of all

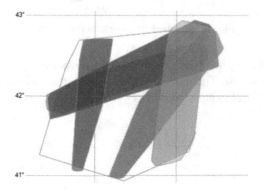

Fig. 5. Clustered standard flows in sector LECMDGU.

of the trajectories in the flow, mean and dispersion of entry and exit flight level ($FL_{in} \pm \sigma_{FLin}$, $FL_{out} \pm \sigma_{FLout}$), Cartesian coordinates and mean and dispersion heading at entry and exit of the airspace volume under analysis ($HG_{in} \pm \sigma_{HGin}$, $HG_{out} \pm \sigma_{HGout}$).

The interaction between two standard flows is defined as the geometrical intersection taking into consideration their mean trajectory and the standard deviation (in coordinates, flight levels and heading). Four types of interaction are identified focusing on the associated cognitive complexity:

- Type 1: Interaction between two cruise flows;
- Type 2: Interaction between a cruise flow and a climbing/descending flow;
- Type 3: Interaction between two flows, both climbing or both descending;
- Type 4: Interaction between two flows, one climbing and the other one descending.

Figure 6 presents these types of standard flows interaction. Level 1 Vertical view shows if flows are in cruise, climbing or descending. Level 2 Horizontal View shows the heading as well.

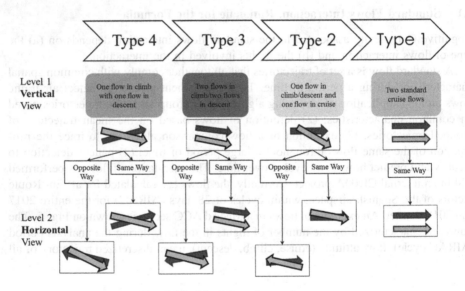

Fig. 6. Standard flows interaction types

The number of flights involved in the standard flows during the interaction is key for complexity assessment. The more flights involved, the more complex the interaction is. The mathematical expression to assess the complexity due to the interaction of two standard flows is:

$$C_{Standard\ Flows\ Interaction} = T_{flow} \times O_{flight}. \tag{1}$$

where T_{flow} is the Interaction Type for Level 1 Vertical View whose weight values are W_i; and O_{flight} is the product of the occupancies of both interacting flows (OCC_1 for Flow 1 and OCC_2 for Flow 2). The Cognitive Complexity increases from Type 1 to Type 4, meaning that $W_1 < W_2 < W_3 < W_4$. For simplification purposes, no weights have been defined for Level 2 Horizontal View Types with one exception for Type 1: if the flows in cruise are in opposite directions, the weight for the interaction is equal to zero, as for safety reasons this type of interaction cannot occur.

4.2 Potential Crossings. Rationale for the Formula

Cognitive Complexity associated to this parameter reflects that potential aircraft encounters in areas not familiar for the ATCo (not classified as critical points category) are more complex. Besides, inspired by [25], the potential crossing may have different severities depending on (1) flight level difference at crossing point; (2) arrival time difference to the crossing point; (3) the average remaining time to the crossing point; (4) the influence of relative vertical speed; and (5) closeness of the potential crossing to a critical point. In order to take into account all the relevant factors, the following formulation is designed.

$$C_{PC} = \sum_{i=1}^{n} (\prod_{j=1}^{5} S_{i,j}). \tag{2}$$

where n is the number of potential crossing: i represents each potential crossing and j considers the five abovementioned factors. Each factor j has an appropriate function S to reflect the corresponding severity of that factor, designed according to (3). The mathematical expression reflects that the severity of a potential crossing increases when the five selected factors decrease.

$$S_{i,j} = e^{(Factor_j/a_j)}. \tag{3}$$

As an assumption, all the factors have the same contribution to the total Potential Crossing complexity. The a_j coefficients are calculated to normalize the functions $S_{i,j}$ in the way that all the factors previously mentioned conduct the same importance. In order to quantify the Potential Crossings complexity it is mandatory to take into account the time horizon and time interval of prediction. Based on the flight transit time in the En-Route sectors, the time horizon is set at 10 min (Fig. 7).

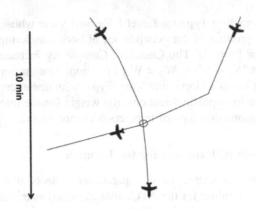

Fig. 7. Flight trajectory prediction (time horizon set at 10 min)

4.3 Flights in Evolution. Rationale for the Formula

The third relevant parameter for the cognitive complexity assessment is the number of climbing/descending flights inside the sector. It is assumed that a flight is in evolution when there is a difference between the planned sector entry and exit Flight Level.

4.4 Flights Out of Standard Flows. Rationale for the Formula

A flight is out of a standard flow when it does not comply with the following criteria:
   In the horizontal plane:

- 90% of the flight trajectory belongs to the flow;
- Heading of the flight at the entry point is between $HG_{in} \pm \sigma_{HGin}$;
- Heading of the flight at the exit point is between $HG_{out} \pm \sigma_{HGout}$.

   In the vertical plane, the limitation for flight assignment is:

- Flight level of the flight at the entry point is inside $FL_{in} \pm \sigma_{FLin}$;
- Flight level of the flight at the exit point is inside $HG_{out} \pm \sigma_{HGout}$.

5 Cognitive Complexity Algorithm

Having characterized and prioritized the different parameters influencing Cognitive Complexity and in order to build the corresponding algorithm, an empirical methodology has been defined based on the data collected in Real Time Simulations (RTS) conducted within the framework of the SESAR 1 programme [10]. During these RTS, four upper and lower sectors were analyzed (LECMDGU, LECMDGL, LECMPAU and LECMPAL) from Madrid ACC. Several ATCos from the Spanish Air Navigation Service Provider, ENAIRE, participated in the simulations.

5.1   Data

The data, which constitute the basis of the study presented, were collected for each validation run throughout the simulations performed and consist of:

- *ISA Values:* ATCo gave subjective MWL by means of Instantaneous Self - Assessment (ISA) values every two minutes in a scale from 1 to 5. The results obtained showed that giving ISA values every two minutes does not interfere with the main task in normal conditions [10]. It should be noted that the ISA values, for being subjective measures, might be different for similar operational conditions depending on the ATCo feeling. Additionally, it was detected that in situations with high loads of traffic, the ISA was randomly provided by the ATCo to avoid interferences with his/her primary tasks. Therefore, in order to reduce the subjectivity of the gathered data, the ISA values were reviewed and corrected by external ATCo who did not participate in the validation exercises. A total number of five scenarios of one-hour duration were reviewed. Considering these five scenarios, and that 25 ISA values are available per run, a total amount of 5 h of simulation and 125 ISA samples could be analyzed.
- *Flight Trajectories:* radar track data was retrieved from the Industrial Business Platform used during the RTS including, for every aircraft, the 4D position every five seconds (time, latitude, longitude and altitude), as well as velocity in the three axis (v_x, v_y, v_z).

In particular, the scenarios selected for each run of the RTS covered both the upper and lower airspace of the Spanish Airspace in Madrid North Area Control Center. For each run, two sectors distributed into four Controller Working Position (CWP) were measured, considering both the Executive and Planner controllers.

5.2   Definition and Calibration

In order to identify the Cognitive Complexity algorithm that best fits with ATCo Subjective MWL (i.e. ISA values), three basic functions that combine the complexity contributing factors described in Sect. 4 (standard flow interactions, flights outside standard flows, flights in evolution, potential crossings), have been defined (4), (5), (6). Based on [26], only non-linear functions have been selected as possible candidates:

$$f_{quadratic} = \sum_{i=1}^{N} \alpha_i + \beta_i x_i + \gamma_i x_i^2. \tag{4}$$

$$f_{exponential} = \sum_{i=1}^{N} \alpha_i + \beta_i e^{\gamma_i x_i}. \tag{5}$$

$$f_{quadradtic-exponential} = \sum_{i=1}^{N} \alpha_i + \beta_i x_i + \gamma_i x_i^2 + \varepsilon_i e^{\delta_i x_i}. \tag{6}$$

with N the number of contributing factors; αi, βi, γi, εi, δi, the calibration coefficients associated to each contributing factor; and xi the value of the contributing factor i. It should be noted that due to the small size of the dataset, big data techniques might not provide the optimum performance. Thus, the methodology described in this section was found as an appropriate alternative. In order to determine the calibration coefficients that provide the complexity algorithm with minimum relative prediction error in comparison with the ISA values, the following methodology was applied:

- As a first step, the values of each contributing factor are associated to the corresponding ISA value based on the corresponding time instant.
- Secondly, the identified values of each contributing factor are introduced in the set of basic functions (4), (5) and (6), computing the associated cognitive complexity value, with the calibration coefficients unknown at this phase.
- Finally, the function coefficients are determined by optimizing the error function defined below between the cognitive complexity value computed with each basic function at every time instant and the respective ISA Fig. (7). For this purpose, and trying to avoid the over-fitting effect, only 80% of the dataset is to be used, leaving the remaining 20% for testing purposes.

$$Error = \sum_{t=1}^{M} |f_t - ISA_t|. \tag{7}$$

In (7), M represents the total number of time instants for which the cognitive complexity value is estimated; and f_t and ISA_t are the computed basic functions and ISA values, respectively, for each time instant t. Once computed, the error function is to be optimized with the Optimization Toolbox provided by Matlab. Particularly, an unconstrained nonlinear optimization method (*fmincon*) has been selected as this algorithm can find the best fitting coefficients to get the minimum error.

5.3 Validation

Having established the calibration coefficients for the multiple functions, and in order to validate them as well as to further characterize the relation between the Cognitive Complexity and the ISA, two types of sequential validation activities have been performed.

1. *Assessment of the Quality of the Prediction Model*

First of all, to determine the quality of the Cognitive Complexity prediction for the selected functions, a Root-Mean-Square Error (RMSE) analysis is to be performed between the predicted Cognitive Complexity and the associated ISA value. For this purpose, the RMSE value is computed for both the training and testing datasets, establishing the comparison between them. This measure, frequently used to characterize the error between the values predicted or estimated by a model and the values observed, can be computed as follows:

$$RMSE = \sqrt{\frac{\sum\limits_{t=1}^{M} (f_t - ISA_t)^2}{M}}. \tag{8}$$

where the numerator corresponds to the squared prediction error, and the denominator is the total number of time instants for which the cognitive complexity is predicted.

Table 3 shows the RMSE computed value for each basic function taking into consideration the calibration coefficients obtained for the model. The range of ISA values predicted were on a continuous scale from 1 to 5. Although for all the functions the testing dataset shows a higher RMSE than the training one, which could indicate that the model is slightly over-fitting the data, in the Quadratic-Exponential approach both errors might be considered as similar.

Table 3. RMSE assessment

Function	Quadratic	Exponential	Quadratic-Exponential
RMSE Training Dataset	0.71	0.72	0.70
RMSE Testing Dataset	0.76	0.82	0.72

Complementary, Figs. 8 and 9, graphically illustrate the relation between the predicted cognitive complexity and the associated ISA value for both the training and testing datasets respectively.

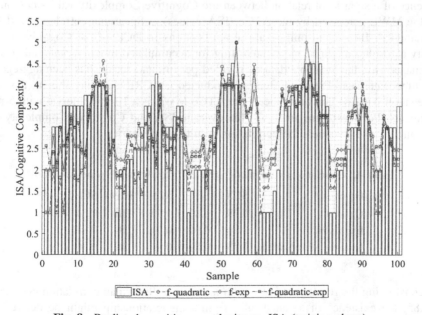

Fig. 8. Predicted cognitive complexity vs. ISA (training phase)

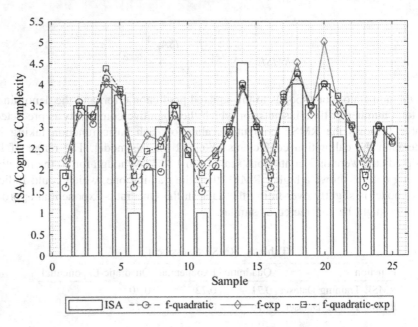

Fig. 9. Predicted cognitive complexity vs. ISA (testing phase)

(1) *Assessment of the Relation between Cognitive Complexity and ATCo MWL (ISA)*

Once the first validation step achieved a positive result, in order to determine the existence of a significant relation between the Cognitive Complexity values computed and the MWL perceived by the ATCo (ISA values), a Spearman correlation test has been applied. Bearing in mind that the applications in DCB of the Cognitive Complexity algorithm might be more interesting for accumulated time intervals rather than instantaneously, the available dataset, divided per run and sector, has been aggregated into time intervals. Previous studies conducted at CRIDA have shown that the aggregation window that best reflects the ATCo MWL is a 10-min interval with 5-min increments. For these time intervals, the average ISA and Cognitive Complexity are calculated. Table 4 gathers the correlation results per scenario.

Table 4. Correlation between cognitive complexity and ISA per scenario

Function	SCN1	SCN2	SCN3	SCN4	SCN5
Quadratic	0.78	0.89	0.93	0.81	0.84
Exponential	0.74	0.90	0.94	0.84	0.81
Quadratic-Exponential	0.73	0.89	0.93	0.82	0.86

Considering the results obtained, with an average Spearman correlation coefficient of 0.85, the existence of a statistically significant relationship might be determined between the defined Cognitive Complexity algorithm and the ATCo MWL. Furthermore, the basic function that best fits is the combination of the quadratic and exponential approach.

6   Cognitive Complexity Interface. Applicability of the Algorithm

It is worth highlighting that the Cognitive Complexity algorithm defined in this study has been integrated into a DCB prototype in the context of the SESAR 2020 Wave 1 Validation Activities. In particular, the algorithm has been integrated into an ad-hoc Demand and Capacity Balancing application created for the execution of Exercise 09.02.03 within "PJ09 – Advanced DCB", in which the Local Traffic Manager actor considered the complexity information for the decision-making relative to the application of Short Term ATFCM Measures (STAM) [22]. Figure 10 illustrates the Human Machine Interface (HMI) for the abovementioned application. The temporal distribution of the Cognitive Complexity as well as the contribution of each complexity factor is displayed as a bar graph in the center of the screen. For each complexity factor defined in the model that contributes to the total Cognitive Complexity value, a different color code has been defined. The list of flights involved in every contributing factor is also presented to identify the eligible flights for the application of DCB measures.

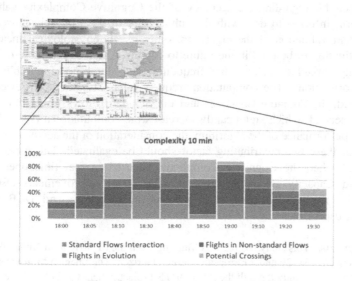

Fig. 10.   Cognitive complexity HMI

7   Conclusions and Further Research

This paper presents the modelling, automation and preliminary calibration of the Cognitive Complexity for its use in the support of ATM planning roles to make decisions on the most suitable DCB measures for solving hotspots. The Cognitive Complexity assessment takes into consideration the ATCo as the center of the system with his/her human limitations. Abstractions are powerful ATCo strategies to avoid the excessive attentional resources demanded from the situation risking the loss of situation awareness and implying wrong decisions that may compromise safe operations. As a hypothesis, the main sources of Cognitive Complexity have been assumed as those traffic and airspace features that do not facilitate the ATCo abstractions. After an analysis and prioritization process, four main parameters have been selected to model Cognitive Complexity: *Standard Flows Interactions, Potential Crossings, Flights in Evolution* and *Flights Out of Standard Flows*. Then, a set of basic functions (quadratic, exponential and quadratic-exponential) defined as a combination of the chosen parameters has been tested. This analysis resulted in the identification of the quadratic-exponential function as the most appropriate approach to represent the Cognitive Complexity as it is the best fitting function with empirical data on subjective ATCo MWL (ISA) gathered during the SESAR 1 RTS. Considering the correlation coefficients obtained with the statistical tests performed, a significant relationship between the Cognitive Complexity values and the ATCo MWL can be established. This does not only reflect the capability of the model presented in this paper to assess complexity, but more importantly to predict it. During the definition, calibration, and validation of the Cognitive Complexity algorithm, a series of limitations have been identified that should be dealt with in further research such as the need to increase the dataset (only 125 ISA samples could be analyzed). Larger datasets would provide significant benefits regarding the accuracy of the Cognitive Complexity values computed. In turn, in order to deal with the subjectivity of the ISA values used for the calibration and validation of the algorithm, other psychophysiological measures are under investigation to be used in the future to capture ATCo MWL more objectively. Eye-Tracking variables such as the frequency and duration of eye-blinks show promising correlations. The computation frequency has also to be reviewed. In the model defined, the Potential Crossings and the Standard Flows Interaction are computed every second, which significantly increases the computational cost. In order to improve the performance of the algorithm, the consideration of the adequate calculation frequency for these two contributing factor should be evaluated. This aspect is especially critical when the Cognitive Complexity is used in real time operations for Demand and Capacity Balancing tasks. Finally, HMI improvements also should be expected, incorporating complexity what-if functionalities and DCB measures management capabilities.

Acknowledgments. This study has received funding from the SESAR Joint Undertaking under grant agreement No 731730 under European Union's Horizon. The Spanish Ministry of Industry has also supported the study through the AIRPORTS Project, where CRIDA and the University of Granada are participants.

References

1. Hilburn, B: Cognitive complexity in air traffic control: a literature review. EEC note 4.04 (2004)
2. Gopher, D., Donchin, E.: Workload: an examination of the concept. In: Boff, K.R., Kaufman, L., Thomas, J.P. (eds.) Handbook of Perception and Performance, vol. 2: Cognitive Processes and Performance, pp. 41–49. Wiley, New York (1986)

3. Mitchell, M.: Complexity: A Guided Tour. Oxford University Press, Oxford (2009)
4. Cañas, J.J., et al.: Mental workload in the explanation of automation effects on ATC performance. In: Longo, L., Leva, M.C. (eds.) H-WORKLOAD 2018. CCIS, vol. 1012, pp. 202–221. Springer, Cham (2019). https://doi.org/10.1007/978-3-030-14273-5_12
5. Netjasov, F., Janić, M., Tošić, V.: Developing a generic metric of terminal airspace traffic complexity. Transportmetrica 7(5), 369–394 (2011)
6. Zhang, M., Shan, L., Zhang, M., Liu, K., Yu, H., Yu, J.: Terminal airspace sector capacity estimation method based on the ATC dynamical model. Kybernetes 45, 884–899 (2016)
7. Tobaruela, G., Schuster, W., Majumdar, A., Ochieng, W.Y., Martinez, L., Hendrickx, P.: A method to estimate air traffic controller mental workload based on traffic clearances. J. Air Transp. Manag. 39, 59–71 (2014)
8. Kontogiannis, T., Malakis, S.: Cognitive Engineering Cognitive Engineering and Safety Organization in Air Traffic Management. CRC Press, Boca Raton (2017)
9. SESAR JU, P04.07.01-D72 Step1 Final Validation Report (2016)
10. SESAR JU, P05.03-D99-Validation Report EXE-05.03-VP-804 (2016)
11. Suárez, N., López, P., Puntero, E., Rodriguez, S: Quantifying air traffic controller mental workload. Fourth SESAR Innovation Days (2014)
12. Cañas, J.J., Ferreira, P., Puntero, E., López, P., López, E., Gómez, F: An air traffic controller psychological model with automation. In: 7th EASN International Conference, Warsaw, Poland (2017)
13. López, P., Puntero, E., López, E., Cañas, J.J., Ferreira, P., Gómez, F., Lucchi, F: Quantitative prediction of automation effects on ATCo Human Performance. In: ICRAT, Barcelona, Spain (2018)
14. Histon, J.M., Hansman, R.J: Mitigating complexity in air traffic control: the role of structure-based abstractions. Report no. ICAT-2008-05. Department of Aeronautics and Astronautics, Massachusetts Institute of Technology, Cambridge, MA (2008)
15. Endsley, M.R.: Toward a theory of situation awareness in dynamic systems. Hum. Fact. J. Hum. Factors Ergon. Soc. 37, 32–64 (1995)
16. Rabinbach, A.: The Human Motor: Energy, Fatigue, and the Origins of Modernity. University of California Press, Berkeley (1990)
17. Kahneman, D.: Attention and Effort. Prentice-Hall, Englewood Cliffs (1973)
18. Wickens, C.D.: Multiple resources and performance prediction. Theor. Issues Ergon. Sci. 3(2), 159–177 (2002)
19. Wickens, C.D., McCarley, J.S.: Applied Attention Theory. CRC Press, Boca Raton (2007)
20. Vicente, K.J.: Cognitive Work Analysis: Toward Safe, Productive, and Healthy Computer-Based Work. Lawrence Erlbaum Associates Publishers, Mahwah (1999)
21. Tyler, S., Neukom, C., Logan, M., Shively, J.: The MIDAS human performance model. In: Proceedings of the Human Factors and Ergonomics Society, 42nd Annual Meeting (1998)
22. SESAR JU, PJ09 OSED, Edition 00.01.12 (2018)
23. Walker, G.H., Stanton, N.A., Salmon, P.M., Jenkins, D.P., Rafferty, L.: Translating concepts of complexity to the field of ergonomics. Ergonomics 53(10), 1175–1186 (2010)
24. Mogford, R.H., Guttman, J.A., Morrow, S.L., Kopardekar, P.: The complexity construct in air traffic control: a review and synthesis of the literature (1995)
25. Zhang, J., Yang, J., Wu, C.: From trees to forest: relational complexity network and workload of air traffic controllers. Ergonomics 58(8), 1320–1336 (2015)
26. MacKay, D.M.: Psychophysics of perceived intensity: a theoretical basis for Fechner's and Stevens' laws. Science 139, 1213–1216 (1963)

EEG-Based Workload Index as a Taxonomic Tool to Evaluate the Similarity of Different Robot-Assisted Surgery Systems

Gianluca Di Flumeri[1,3(✉)], Pietro Aricò[1,3], Gianluca Borghini[1,3],
Nicolina Sciaraffa[2,3], Vincenzo Ronca[3], Alessia Vozzi[3],
Silvia Francesca Storti[5], Gloria Menegaz[5], Paolo Fiorini[5],
and Fabio Babiloni[1,3,4]

[1] Department of Molecular Medicine, Sapienza University of Rome, Rome, Italy
{gianluca.diflumeri,pietro.arico,gianluca.borghini,
fabio.babiloni}@uniroma1.it
[2] Department of Anatomical, Histological, Forensic and Orthopedic Sciences,
Sapienza University of Rome, Rome, Italy
nicolina.sciaraffa@uniroma1.it
[3] BrainSigns Srl, Rome, Italy
[4] Department of Computer Science, Hangzhou Dianzi University,
Hangzhou, China
[5] Department of Computer Science, University of Verona, Verona, Italy
{silviafrancesca.storti,gloria.menegaz,
paolo.fiorini}@univr.it

Abstract. In operational fields, there is a growing use of simulators during training protocols because of their versatility, the possibility of limiting costs and increasing efficiency. This work aimed at proposing an EEG-based neurometric of mental workload, previously validated in other contexts, as a taxonomic tool to evaluate the similarity, in terms of cognitive demands, of two different systems: the da Vinci surgical system, leader in the field of robotic surgery, and the Actaeon Console by BBZ, basically a cheaper simulator aimed to train students to use the da Vinci system. Such a neurophysiologic evaluation of the workload demand was also integrated by information derived by the task performance and self-reports. The results validated the proposed EEG-based workload index and indicated the potentially fruitful use of simulators because of their high similarity in terms of cognitive demands.

Keywords: Workload · EEG · Machine-learning · Neurometrics · Robot-assisted surgery

1 Introduction

Thanks to the recent technological progress, driven by the great interest and appeal raised by Virtual Realities, simulators are becoming more and more realistic. Not surprisingly, their employment has been promoted across different domains, firstly in aviation [1], but also in maritime domain [2], in the military one [3] and in different

© Springer Nature Switzerland AG 2019
L. Longo and M. C. Leva (Eds.): H-WORKLOAD 2019, CCIS 1107, pp. 105–117, 2019.
https://doi.org/10.1007/978-3-030-32423-0_7

applications of medical service, such as laparoscopic [4] and orthopaedic [5] surgery. In all these fields, the use of simulators is particularly appreciated within the training protocols, since they allow to simulate a potentially infinite number of situations and repetitions, even situations that are risky to reproduce in reality, but at the same time saving money and time [6]. Let us think for example to the aviation domain: the pilots' training is mainly performed at simulators, since the costs of having/renting aircrafts and related facilities, as well as the fuel and the maintenance, would be unsustainable for a training center [6]. The healthcare domain has been also run over by this phenomenon, i.e. the use of simulators, especially in the field of laparoscopic and robot-assisted surgery, where the medical doctor's activity already implies interaction with robotic devices, thus with high similarity between simulated and real conditions [4]. In this application, the leading robotic system is the *da Vinci* system (Intuitive Surgical, Sunnyvale, CA, USA), able to support almost all the operations of laparoscopic surgery [7]. Since its introduction in 1998, robot-assisted laparoscopic surgery has seen rapid widespread adoption. As more hospitals and surgical specialties use the *da Vinci* system, there has been increased demand for the development of a validated training curriculum similar to the *Fundamentals of Laparoscopic Surgery* (FLS) program [8]. The correct use of the robot would improve surgeon's ability and efficiency, however, the robotic procedure may stress the surgeon and the operating room team and increase the risk for the patients. Over the past decade, there was increased interest in the measurement of surgical skills and performance in laparoscopy, especially as these relate to the evaluation of training [9]. Expertise of a surgeon results from a complex process of learning and practicing with development of networks of implicit knowledge, technical skills, and memory [10, 11]. Traditional training assessment methods rely on the supervision of the trainees during repeated practice. Although the use of performance efficiency measures (speed, movement economy, errors) and ergonomic assessments are relatively well established, the evaluation of the cognitive engagement of the trainees is more rare [12]. Cognition-based assessment is critical to grade from novice to expert status during the course of training. The related information is traditionally gathered using self-reported questionnaires, such as the *NASA – Task Load Index* (NASA-TLX) [13]. The major drawback of these measures is that they cannot be unobtrusively administered during task execution, but are assessed at the end of the task, which compromises the accuracy and reliability of the measurement itself [14, 15]. Moreover, such methods suffer from subjective biases due to self-assessment, as well as from operator–dependent judgments. Neurophysiological measures, such as the *Electroencephalogram* signal (EEG), allow to objectively assess the cognitive state under which the user is performing the considered task. The effectiveness of this approach was already explored in a variety of applications ranging from human-robot interaction to tasks involving motor-cognitive task and air-traffic-control [16]. The use of neurophysiological methodologies has been already employed to evaluate the cognitive performance during training protocols [17, 18], including those ones based on laparoscopic simulators [19]. However, the actual similarity between a device and the related simulator (in terms of cognitive demands) is still unclear. Undoubtedly, two systems would be compared considering different factors, including both cognitive physical and practical aspects, anyhow cognitive demands are one of the main variables affecting performance of humans while interacting with interfaces [20, 21]. This work

aimed at proposing an EEG-based neurometric of mental workload, previously validated in other contexts [22–24], as a taxonomic tool to evaluate the similarity, in terms
of cognitive demands, of two different systems: the *da Vinci* system itself, leader in the
field of robotic surgery, and the Actaeon Console (BBZ srl, Italy), basically a cheaper
simulator aimed to train students to use the da Vinci system. Such a neurophysiologic
evaluation of the workload demand was also integrated by information derived by the
task performance and self-reports.

2   Materials and Methods

2.1   The Experimental Protocol

Twenty healthy students (half male) from the Faculty of Medicine of the University of
Verona have been recruited, on a voluntary basis, in this study. They were selected in
order to have a homogeneous experimental group in terms of age and expertise. The
experiment was conducted following the principles outlined in the Declaration of
Helsinki of 1975, as revised in 2000. Informed consent was obtained from each subject
on paper, after the explanation of the study. The experimental protocol consisted in
executing some laparoscopy-like tasks by using two different systems for the robot-
assisted surgery: the *da Vinci Surgical system* (Intuitive Surgical Inc, USA), leader in
the field of robotic surgery, and the *Actaeon Console* (BBZ srl, Italy), the low-cost
simulator of the da Vinci system (Fig. 1).

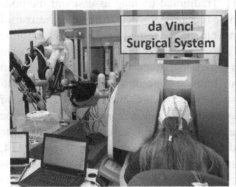

Fig. 1. The same student during the experiments performing the experimental tasks by the *da
Vinci Surgical System* (on the left) and by the *BBZ Actaeon Simulator* (on the right).

In particular, after a training period of 3 weeks aimed at making the subjects
confident with both the systems, they had to perform two Easy and two Hard repetitions of the TornWire task. Such a task, generally employed to train the subjects to use
this kind of devices, consisted in bringing a ring along a twisted wire (the wire passed
through the hole of the ring), before in a way and after in the contrary one, in a
maximum time of four minutes, avoiding any contact between the ring and the wire

itself. Each contact was considered by the system as error. The difficulty of the task depended on the severity of the wire twisting. The execution order of the four tasks (two easy and two hard TornWire tasks per subject) was randomized across the subjects to avoid any habituation and learning effect [17].

2.2   Data Recording

During the whole experimental protocol, including a one-minute-long "Closed Eyes" condition preceding the experimental tasks, the EEG signals were recorded using the digital monitoring BEmicro system (EBNeuro, Italy). Twelve EEG channels (FPz, AF3, AF4, F3, Fz, F4, P3, P7, Pz, P4, P8 and POz), placed according to the 10–20 International System, were collected with a sampling frequency of 256 Hz, all referenced to both the earlobes, grounded to the Cz site, and with the impedances kept below 20 kΩ.

The recoded EEG signal was firstly band-pass filtered with a fifth-order Butterworth filter (high-pass filter cut-off frequency: 1 Hz, low-pass filter cut-off frequency: 30 Hz). The Fpz channel was used to remove eyes-blink contributions from each channel of the EEG signal, by using the REBLINCA algorithm [25]. This step is necessary because the eyes-blink contribution could affect the frequency bands correlated to the mental workload, in particular the theta EEG band. For other sources of artefacts (external interferences as well as movements performed by the user during their operative activity), specific procedures of the EEGLAB toolbox were used [26]. Firstly, the EEG signal was segmented into epochs of 2 s (Epoch length), shifted of 0.125 s (Shift). This windowing was chosen to satisfy the trade-off between having a high number of observations, in comparison with the number of variables, and respecting the stationarity conditions of the EEG signal [27]. In fact, this is a necessary hypothesis in order to proceed with the spectral analysis of the signal. The EEG epochs with a signal amplitude exceeding ± 100 μV (*Threshold criterion*) was marked as "artefact". Then, each EEG epoch was interpolated in order to check the related slope of the trend (*Trend estimation*). If such slope was higher than 20 $\mu V/s$, the considered epoch was marked as "artefact". The last criterion was based on the EEG Sample-to-sample difference (*Sample-to-sample criterion*). If such difference, in terms of absolute amplitude, is higher than 25 (μV), i.e. an abrupt variation (no-physiological) happened, the EEG epoch was marked as "artefact". At the end, the EEG epochs marked as "artefact" were removed from the EEG dataset with the aim of obtaining a clean EEG signal from which to estimate the brain parameters for the different analyses. All the previous mentioned values were chosen following the guidelines suggested by Delorme and Makeig [26]. From the clean EEG dataset, the Power Spectral Density (PSD) was calculated for each EEG channel for each epoch using a Hanning window of the same length of the considered epoch (2 s length, that means 0.5 Hz of frequency resolution). The application of a Hanning window helps smoothing the contribution of the signal close to the end of the segment (Epoch), improving the accuracy of the PSD estimation [28]. Then, the EEG frequency bands of interest were defined for each user by the estimation of the *Individual Alpha Frequency* (IAF) value [29]. In order to have a precise estimation of the alpha peak and, hence of the IAF, the users were asked to keep the eyes closed for a minute before starting the experimental tasks: the EEG data from parietal sites (where the alpha synchronization is higher) during this experimental condition were then used for the IAF estimation. Finally, a spectral features matrix (EEG channels × Frequency bins) was obtained in the frequency bands directly correlated to the mental workload. In particular, only the theta band [IAF − 6 Hz ÷ IAF − 2 Hz], over the EEG frontal channels, and the alpha band [IAF − 2 Hz ÷ IAF + 2 Hz], over the EEG parietal channels, have been considered as variables for the mental workload evaluation, according to the evidences from scientific literature and previous studies [30]. At this point, the features domain is ready to be treated by applying machine learning techniques. In particular, the automatic-stop-StepWise Linear Discriminant Analysis (asSWLDA) [22] was used to the purpose. On the basis of the training dataset, the asSWLDA selects the most relevant spectral features to discriminate the Mental Workload of the subjects within the different experimental conditions (i.e. for example EASY and HARD, or in general the two extremes of the mental state to evaluate). Once identified such spectral features, the asSWLDA assigns to each one specific weights ($w_{i\ train}$), plus a bias (b_{train}), such that the asSWLDA discriminant function ($y_{train}(t)$) takes the value 1 in the hardest condition and 0 in the easiest one. This step represents the *Training phase* of the classifier. Later on, the weights and the bias as determined during the training phase are used to calculate the Linear Discriminant function ($y_{test}(t)$) over the testing dataset (*Testing phase*). Finally, a moving average of 8 s (8MA) was applied to the $y_{test}(t)$ function in order to smooth it out by reducing the variance of the measures. Hereinafter, the output is named *EEG-based Workload index (W_{EEG})*. Here below are reported the training asSWLDA model (Eq. 1, where $f_{i\ train}(t)$ represents the PSD matrix of the training dataset for the data window for the time sample t, and of the ith feature), the testing asSWLDA Discriminant Function (Eq. 2, where $f_{i\ test}(t)$ is defined as $f_{i\ train}(t)$ but related to the testing dataset) and the equation of the *EEG-based workload index, W_{EEG}* (Eq. 3).

$$y_{train}(t) = \sum_i w_{itrain} \cdot f_{itrain}(t) + b_{train} \tag{1}$$

$$y_{test}(t) = \sum_i w_{itrain} \cdot f_{itest}(t) + b_{train} \tag{2}$$

$$W_{EEG} = 8MA(y_{test}(t)) \tag{3}$$

In addition, the subjective measures, in terms of perceived workload, have been collected. At the end of each execution, the subjects were asked to fill the NASA–TLX questionnaire [13], with the aim of collecting data about the perception of the workload across the training sessions. The total workload score was calculated as combination of six factors (Mental Demand, Physical Demand, Temporal Demand, Performance, Effort and Frustration), and it ranges from 0 to 100. Finally, during the execution of the task, each system collected several parameters related to the control of the robots and correctness of the activities, taking into account also the time of execution, the task completion percentage and the number of errors, and produced a performance index from 0% (the worst score) to 100% (the best score).

The Performed Analysis

The analyses have been organized in two sections, in order to answer the two experimental questions: the validation of the workload evaluation algorithm, and its application for the comparison of the two robotic systems, i.e. the Da Vinci robot and the BBZ simulator.

Validation of the EEG-Based Workload Evaluation Algorithm. Each subject, with each system, performed two Easy and two Hard TornWire tasks. For each system, the asSWLDA algorithm has been trained with a couple of Easy-Hard tasks and tested on the remaining ones. Also, in order to test the robustness of the algorithm towards the system used (i.e. the algorithm is sensitive to the task difficulty, or better said the mental workload, independently from the system), the algorithm has been trained by crossing the training and testing dataset from the two systems: it was trained on a couple of Easy-Hard tasks performed with Da Vinci robot and tested on BBZ data, and vice-versa (labelled "Mixed" in the results). A repeated measures ANOVA has been performed on the Area Under Curve (AUC [31]) values of the classifier ROC curve obtained in the three cases (Da Vinci, BBZ and Mixed). In particular, AUC represents a widely used performance index of a binary classifier: the classification performance can be considered good with an AUC higher than at least 0.7 [32]. Also, to validate the algorithm performance, they have been compared, by three two-tailed paired Student's T-tests (one for each case), with the performance obtained by employing the algorithm on the testing dataset randomly shuffled, in order to induce misclassification.

Comparison of Different Systems (BBZ vs Da Vinci). Once validated the algorithm for the estimation of the EEG-based Workload index, the neurophysiological scores have been used to explore the possibility to employ them for comparing the two systems. Therefore, as well as with the subjective (NASA-TLX) and behavioural (performance) measures, the EEG-based workload scores obtained from the use of the two systems have been compared by three two-tailed paired Student's t-tests, one for each kind of measures (i.e. subjective, behavioural and neurophysiological).

3   Results

Validation of EEG-Based Workload Evaluation

In order to validate the algorithm for the mental workload estimation in terms of robustness toward the system used by the operator, it has been applied in three ways: in the first two cases, it has been used in the "right" way by using training and testing data from the same system (BBZ simulator and Da Vinci robot). Thirdly, it has been applied also crossing the dataset, i.e. training on the BBZ and testing on Da Vinci data, and vice-versa. In addition, for all the cases the algorithm has been applied randomizing the testing data. The ANOVAs analysis (Fig. 2) revealed no significant differences in terms of algorithm performance independently from the system used: for all the three cases (BBZ, Da Vinci and mixed, blue line in figure) the classification discriminability, measured through AUC values, in distinguishing the Easy from the Hard TornWire task achieved values in the range between 70% and 75%. In addition, all the three 2-tailed paired T-tests (one for each case: BBZ, Da Vinci and Mixed) have shown significant differences ($p < 10^{-6}$) with respect to the application on randomized testing data (red line in Fig. 2).

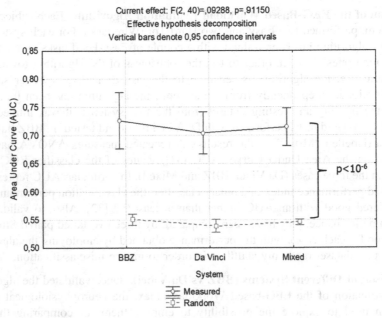

Fig. 2. ANOVAs analysis on the AUC values obtained by training and testing the algorithm on the BBZ simulator, on the Da Vinci robot, and crossing the datasets, i.e. training on BBZ and testing on the Da Vinci and vice-versa. Also, the reported significance is related to the three two-tailed paired T-tests performed for each case (BBZ, Da Vinci and Mixed), between the algorithm correct performance ("Measured", blue line) and those ones obtained randomizing the testing data ("Random", red line). (Color figure online)

Comparison of Different Systems (BBZ vs Da Vinci)

Once validated the method, the EEG-based workload indexes have been used to compare the two systems, i.e. the BBZ simulator and the Da Vinci robot, from a cognitive point of view. Figures 3 and 4 show the results of the two-tailed paired T-tests respectively on the behavioural (i.e. performance) and subjective (i.e. NASA-TLX) measures of workload. In both the cases, no significant differences could be detected by the analysis (respectively $p = 0.96$ and $p = 0.31$).

The EEG-based workload indexes confirmed these results: the two-tailed paired T-test (Fig. 5) did not reveal any significant differences between the two systems ($p = 0.66$). In addition, the Pearson's correlation analysis (in Fig. 6 the scatterplot of such analysis) highlighted a positive and significant correlation between the EEG-based workload indexes as obtained on the same subjects managing the two systems. In other words, these results confirm the high similarity of the two systems in terms of cognitive demand.

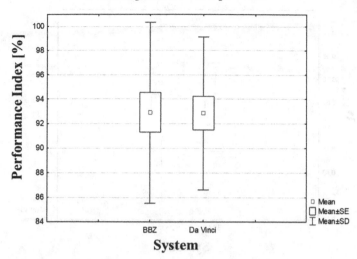

Fig. 3. Two-tailed paired T-test on the performance indexes between the two systems, i.e. the BBZ simulator and the Da Vinci robot.

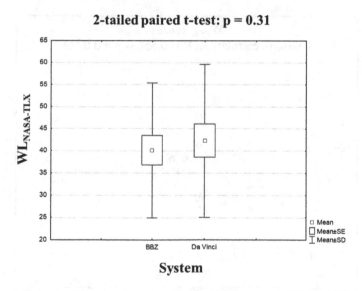

Fig. 4. Two-tailed paired T-test on the subjective workload measures (NASA-TLX) between the two systems, i.e. the BBZ simulator and the Da Vinci robot.

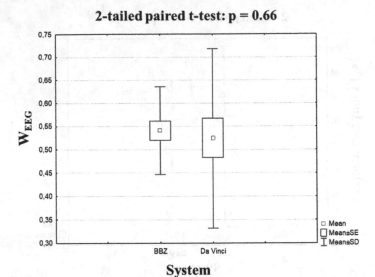

Fig. 5. Two-tailed paired t-test on the EEG-based Mental Workload indexes (W_{EEG}) between the two systems, i.e. the BBZ simulator and the Da Vinci robot.

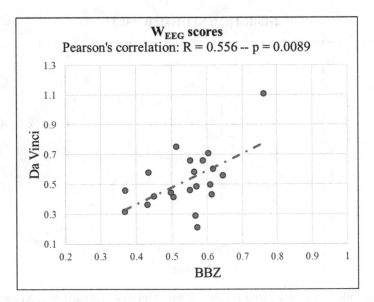

Fig. 6. Scatterplot highlighting the positive and significant correlation between EEG-based Mental Workload indexes obtained for each subject (1 dot = 1 subject) managing the two systems, i.e. the BBZ simulator and the Da Vinci robot. In red the tendency line. (Color figure online)

4 Discussion

Also in the medical domain, the relationship between human errors and performance impairment due to a high mental workload has been deeply investigated and documented [33]. With the exploitation of robotic technology, more and more robotic systems, in particularly in surgery, have been developed with the aim to support the surgeon. In this experiment, twenty healthy students of Medicine of the University of Verona were recruited. They had to perform similar laparoscopy-like tasks, under different levels of difficulty (Easy and Hard), by using two different systems for the robot-assisted surgery: the da Vinci Surgical system (Intuitive Surgical Inc, USA), leader in the field of robotic surgery, and the Actaeon Console (BBZ srl, Italy), basically a cheaper simulator aimed to train students to use the da Vinci system. Firstly, a machine-learning approach consisting in the use of the asSWLDA algorithm [22], was employed and tested in terms of reliability in measuring the user's mental workload from his/her brain activity, measured through EEG. The results in terms of classes (Low and High workload) discriminability have demonstrated the reliability of the EEG-based workload evaluation method also in such a kind of application. In fact, this algorithm had been previously validated in the aviation [22, 23] and automotive [24] domains, however the validation in this kind of Human-Machine Interaction was still missing. In the present study, the AUC values related to the classifier performance in distinguishing the Easy and Hard conditions of the same laparoscopic-like tasks were all higher than 0.7 (considered as a reference threshold for good performance [32]) and up to 0.8. Also, they were significantly higher ($p < 10^{-6}$) than the performance obtained after randomizing the data classes (i.e. random classifier). Furthermore, and even more interestingly, the robustness of the algorithm has been proven by the fact that also calibrating the method on one system (e.g. the BBZ) and testing it on the other (e.g. the Da Vinci), the performance in terms of workload classification remained stable, i.e. without any significant difference. This provides evidence in support of the fact that the method is sensitive to the probed cognitive mechanism, i.e. the Mental Workload, and not to spurious actions that depend from the used system. In fact, the contrasted tasks were similar and they differ only in terms of difficulty. This result paves the way for several new applications of these methodologies, since (i) from one side, it provides an indirect tool to measure the similarity of different devices while performing the same task, and (ii) from the other side it also provide evidences about the feasibility of cross-task calibrations, a still open issue in literature [34].

Last but not least, this experiment demonstrated that it is also possible to use these EEG-based Workload indexes to compare different devices from a cognitive point of view. In fact, according to what arose from the analysis of behavioural (i.e. performance) and subjective (self-assessed evaluation) measures, the two systems were not significantly different ($p = 0.66$) in terms of mental workload experienced by the user (measured through EEG), and even more the scores related to the two systems across the subjects resulted significantly ($R = 0.56$; $p = 0.009$) positively correlated. Therefore, the use of such a neurophysiological approach could provide a complementary tool to be employed in the recent field of neuroergonomics [35], enabling the integration of traditional methods, such as questionnaires and behavioural analysis, with neurophysiological measures, such as EEG, while evaluating the impact on the mental workload of different systems designed for the same activity. Undoubtedly, devices and their simulators should be compared by considered several other factors apart from the workload, however cognitive demands are one of the main variables affecting performance of humans while interacting with interfaces. Therefore, in this scenario the present study does not aim at providing the unique measure (i.e. the EEG-based workload one) to compare devices, but at validating innovative and reliable approaches to integrate the traditional ones to obtain higher levels of reliability and larger perspectives over the problem. In addition, thanks to the recent success of wearable technology and the enhanced reliability of innovative Brain-Computer Interfaces [36, 37], these results suggest also future applications employing such a kind of approach to support in real time the surgeon and the operators by making them aware of their current mental workload, even while working in team [38, 39]. It has to be considered that the use of a couple of similar tasks, differing only in terms of cognitive demand, has to be intended as a cautious way to separate workload variation from any other external variable (i.e. physical movements, type of actions to plan) that would impact mental activity if comparing different tasks. Nevertheless, because of that the conclusions of the present manuscript have to be intended as limited to the proposed case study. Additional experiments involving different types of task are needed to validate the proposed methodology in a larger extent, also including an analysis of variability of such neurometrics employing machine-learning techniques with respect of traditional methods based on subjective measures. For this kind of analysis, larger experimental samples are undoubtedly necessary. In conclusion, the interesting results promote further investigation in order to test the method on larger subjects' samples, including different experimental tasks and even different kinds of devices.

5   Conclusion

Technical skill assessment and training is a critical component in critical scenarios such as in robotic-assisted surgery. The use of simulators would support training protocols, allowing at the same time to reduce costs and improve efficiency. The results of the study demonstrate how neurophysiological measures of mental workload can enhance standard methodologies while evaluating different devices, by providing objective information about cognitive resources employed by the user while interacting with them. Also, the present study provided evidence about the cognitive equivalence between the reference device and its simulator.

Acknowledgements. The BBZ team and the University of Verona are sincerely acknowledged for allowing the use of their facilities.

References

1. Flight Simulation: Virtual Environments in Aviation, 1st edn. (Hardback). Routledge. Routledge.com.   https://www.routledge.com/Flight-Simulation-Virtual-Environments-in-Aviation-1st-Edition/Lee/p/book/9780754642879. Accessed 04 July 2019

2. Sellberg, C., Lindmark, O., Rystedt, H.: Learning to navigate: the centrality of instructions and assessments for developing students' professional competencies in simulator-based training. WMU J. Marit. Aff. **17**(2), 249–265 (2018)

3. Rech, M., Bos, D., Jenkings, K.N., Williams, A., Woodward, R.: Geography, military geography, and critical military studies. Crit. Mil. Stud. **1**(1), 47–60 (2015)

4. Yiannakopoulou, E., Nikiteas, N., Perrea, D., Tsigris, C.: Virtual reality simulators and training in laparoscopic surgery. Int. J. Surg. Lond. Engl. **13**, 60–64 (2015)

5. Vaughan, N., Dubey, V.N., Wainwright, T.W., Middleton, R.G.: A review of virtual reality based training simulators for orthopaedic surgery. Med. Eng. Phys. **38**(2), 59–71 (2016)

6. Andrews, D.H.: Relationships among simulators, training devices, and learning: a behavioral view. Educ. Technol. **28**(1), 48–54 (1988)

7. Maeso, S., et al.: Efficacy of the Da Vinci surgical system in abdominal surgery compared with that of laparoscopy: a systematic review and meta-analysis. Ann. Surg. **252**(2), 254–262 (2010)

8. Ritter, E.M., Scott, D.J.: Design of a proficiency-based skills training curriculum for the fundamentals of laparoscopic surgery. Surg. Innov. **14**(2), 107–112 (2007)

9. Hussein, A.A., et al.: Technical mentorship during robot-assisted surgery: a cognitive analysis. BJU Int. **118**(3), 429–436 (2016)

10. McLeod, P.J., Steinert, Y., Meagher, T., Schuwirth, L., Tabatabai, D., McLeod, A.H.: The acquisition of tacit knowledge in medical education: learning by doing. Med. Educ. **40**(2), 146–149 (2006)

11. Borghini, G., et al.: EEG-based cognitive control behaviour assessment: an ecological study with professional air traffic controllers. Sci. Rep. **7**(1), 547 (2017)

12. Byrne, A.: The effect of education and training on mental workload in medical education. In: Longo, L., Leva, M.C. (eds.) H-WORKLOAD 2018. CCIS, vol. 1012, pp. 258–266. Springer, Cham (2019). https://doi.org/10.1007/978-3-030-14273-5_15

13. Hart, S.G., Staveland, L.E.: Development of NASA-TLX (Task Load Index): results of empirical and theoretical research. In: Hancock, P.A., Meshkati, N. (eds.) Advances in Psychology, vol. 52, pp. 139–183. North-Holland, Amsterdam (1988)

14. Aricò, P., Borghini, G., Di Flumeri, G., Sciaraffa, N., Colosimo, A., Babiloni, F.: Passive BCI in operational environments: insights, recent advances, and future trends. IEEE Trans. Biomed. Eng. **64**(7), 1431–1436 (2017)

15. Moustafa, K., Luz, S., Longo, L.: Assessment of mental workload: a comparison of machine learning methods and subjective assessment techniques. In: Longo, L., Leva, M.C. (eds.) H-WORKLOAD 2017. CCIS, vol. 726, pp. 30–50. Springer, Cham (2017). https://doi.org/10.1007/978-3-319-61061-0_3

16. Borghini, G., et al.: Quantitative assessment of the training improvement in a motor-cognitive task by using EEG, ECG and EOG Signals. Brain Topogr. **29**(1), 149–161 (2016)

17. Borghini, G., et al.: A new perspective for the training assessment: machine learning-based neurometric for augmented user's evaluation. Front. Neurosci. **11**, 325 (2017)

18. Aricò, P., et al.: Human factors and neurophysiological metrics in air traffic control: a critical review. IEEE Rev. Biomed. Eng. **10**, 250–263 (2017)

19. Borghini, G., et al.: Neurophysiological measures for users' training objective assessment during simulated robot-assisted laparoscopic surgery. In: 2016 38th Annual International Conference of the IEEE Engineering in Medicine and Biology Society (EMBC), pp. 981–984 (2016)

20. Wickens, Christopher D.: Mental workload: assessment, prediction and consequences. In: Longo, L., Leva, M.C. (eds.) H-WORKLOAD 2017. CCIS, vol. 726, pp. 18–29. Springer, Cham (2017). https://doi.org/10.1007/978-3-319-61061-0_2

21. Parasuraman, R., McKinley, R.A.: Using noninvasive brain stimulation to accelerate learning and enhance human performance. Hum. Factors J. Hum. Factors Ergon. Soc. **56**(5), 816–824 (2014)
22. Aricò, P., Borghini, G., Di Flumeri, G., Colosimo, A., Pozzi, S., Babiloni, F.: A passive brain-computer interface application for the mental workload assessment on professional air traffic controllers during realistic air traffic control tasks. Prog. Brain Res. **228**, 295–328 (2016)
23. Aricò, P., et al.: Adaptive automation triggered by eeg-based mental workload index: a passive brain-computer interface application in realistic air traffic control environment. Front. Hum. Neurosci. **10**, 539 (2016)
24. Di Flumeri, G., et al.: EEG-based mental workload neurometric to evaluate the impact of different traffic and road conditions in real driving settings. Front. Hum. Neurosci. **12**, 509 (2018)
25. Di Flumeri, G., Aricò, P., Borghini, G., Colosimo, A., Babiloni, F.: A new regression-based method for the eye blinks artifacts correction in the EEG signal, without using any EOG channel. In: Conference Proceedings of Annual International Conference of the IEEE Engineering in Medicine and Biology Society (2016)
26. Delorme, A., Makeig, S.: EEGLAB: an open source toolbox for analysis of single-trial EEG dynamics including independent component analysis. J. Neurosci. Methods **134**(1), 9–21 (2004)
27. Elul, R.: Gaussian behavior of the electroencephalogram: changes during performance of mental task. Science **164**(3877), 328–331 (1969)
28. Harris, F.J.: On the use of windows for harmonic analysis with the discrete Fourier transform. Proc. IEEE **66**(1), 51–83 (1978)
29. Klimesch, W.: EEG alpha and theta oscillations reflect cognitive and memory performance: a review and analysis. Brain Res. Rev. **29**(2–3), 169–195 (1999)
30. Borghini, G., Astolfi, L., Vecchiato, G., Mattia, D., Babiloni, F.: Measuring neurophysiological signals in aircraft pilots and car drivers for the assessment of mental workload, fatigue and drowsiness. Neurosci. Biobehav. Rev. **44**, 58–75 (2014)
31. Bamber, D.: The area above the ordinal dominance graph and the area below the receiver operating characteristic graph. J. Math. Psychol. **12**(4), 387–415 (1975)
32. Fawcett, T.: An introduction to ROC analysis. Pattern Recogn. Lett. **27**(8), 861–874 (2006)
33. Helmreich, R.L.: On error management: lessons from aviation. BMJ **320**(7237), 781–785 (2000)
34. Walter, C., Schmidt, S., Rosenstiel, W., Gerjets, P., Bogdan, M.: Using cross-task classification for classifying workload levels in complex learning tasks. In: 2013 Humaine Association Conference on Affective Computing and Intelligent Interaction, pp. 876–881 (2013)
35. Aricò, P., Borghini, G., Flumeri, G.D., Sciaraffa, N., Babiloni, F.: Passive BCI beyond the lab: current trends and future directions. Physiol. Meas. **39**(8), 08TR02 (2018)
36. Parasuraman, R.: Neuroergonomics: research and practice. Theor. Issues Ergon. Sci. **4**(1–2), 5–20 (2003)
37. Di Flumeri, G., Aricò, P., Borghini, G., Sciaraffa, N., Di Florio, A., Babiloni, F.: The dry revolution: evaluation of three different EEG dry electrode types in terms of signal spectral features, mental states classification and usability. Sensors **19**(6), 1365 (2019)
38. Sciaraffa, N., et al.: Brain interaction during cooperation: evaluating local properties of multiple-brain network. Brain Sci. **7**(7), 90 (2017)
39. Antonacci, Y., Toppi, J., Caschera, S., Anzolin, A., Mattia, D., Astolfi, L.: Estimating brain connectivity when few data points are available: Perspectives and limitations. In: 2017 39th Annual International Conference of the IEEE Engineering in Medicine and Biology Society (EMBC), pp. 4351–4354 (2017)

Applications

Deep Learning for Automatic EEG Feature Extraction: An Application in Drivers' Mental Workload Classification

Mir Riyanul Islam[1]([envelope]), Shaibal Barua[1], Mobyen Uddin Ahmed[1], Shahina Begum[1], and Gianluca Di Flumeri[2,3]

[1] School of Innovation, Design and Engineering, Mälardalen University, Västerås, Sweden
{mir.riyanul.islam,shaibal.barua,mobyen.ahmed,shahina.begum}@mdh.se
[2] BrainSigns s.r.l., Rome, Italy
gianluca.diflumeri@brainsigns.com
[3] Department of Molecular Medicine, Sapienza University of Rome, Rome, Italy

Abstract. In the pursuit of reducing traffic accidents, drivers' mental workload (MWL) has been considered as one of the vital aspects. To measure MWL in different driving situations Electroencephalography (EEG) of the drivers has been studied intensely. However, in the literature, mostly, manual analytic methods are applied to extract and select features from the EEG signals to quantify drivers' MWL. Nevertheless, the amount of time and effort required to perform prevailing feature extraction techniques leverage the need for automated feature extraction techniques. This work investigates deep learning (DL) algorithm to extract and select features from the EEG signals during naturalistic driving situations. Here, to compare the DL based and traditional feature extraction techniques, a number of classifiers have been deployed. Results have shown that the highest value of area under the curve of the receiver operating characteristic (AUC-ROC) is 0.94, achieved using the features extracted by convolutional neural network autoencoder (CNN-AE) and support vector machine. Whereas, using the features extracted by the traditional method, the highest value of AUC-ROC is 0.78 with the multi-layer perceptron. Thus, the outcome of this study shows that the automatic feature extraction techniques based on CNN-AE can outperform the manual techniques in terms of classification accuracy.

Keywords: Autoencoder · Convolutional neural networks · Electroencephalography · Feature extraction · Mental workload

1 Introduction

Driver's mental workload (MWL) plays a crucial role on the driving performance. Due to excessive MWL, drivers undergo a complex state of fatigue which manifests lack of alertness and reduces performance [1]. Consequently, drivers are

© Springer Nature Switzerland AG 2019
L. Longo and M. C. Leva (Eds.): H-WORKLOAD 2019, CCIS 1107, pp. 121–135, 2019.
https://doi.org/10.1007/978-3-030-32423-0_8

prone to committing more mistakes due to increased MWL. It has been revealed that human error is the prime cause of around 72% road accidents per year [2]. So, increased MWL of drivers during driving can produce errors leading to fatal accidents. Driving is a complex and dynamic activity involving secondary tasks i.e. simultaneous cognitive, visual and spatial tasks. Diverse secondary tasks along with natural driving in addition to different road environments increase the MWL of drivers which lead to errors in traffic situations [3]. The alarming number of traffic accidents due to increased MWL leverages the need of determining drivers' MWL efficiently. Several research works have identified mechanisms to measure drivers' MWL while driving both in simulated and real environments [1,4,5]. Methods of measuring MWL can be clustered into three main classes; (i) subjective measures i.e. NASA Task Load Index (NASA-TLX), workload profile (WP) etc., (ii) task performance measures e.g. time to complete a task, reaction time to secondary task etc. and (iii) physiological measures e.g. electroencephalogram (EEG), heart rate measures etc. [6]. The latter, with respect to traditional subjective measures, are intrinsically objective and can be gathered along with the task without asking any additional action to the user. Also, with respect to performance measures, physiological measures do not require as well secondary tasks and are generally able to predict a mental impairment, while on the contrary performance generally degrades when the user is already overloaded [7–9]. Due to the vast availability of measuring technology, portability and capability of indicating neural activation clearly, the major concern of this work is the physiological measures, specifically, EEG. With the increase of data storage and computation power data-driven machine learning (ML) techniques have been becoming popular means of quantifying MWL from EEG signals.

Relevant features extracted from the EEG signals are the sine qua nons for quantifying MWL. Currently, feature extraction is done using theory driven manual analytic methods that demand huge time and effort [10,11]. The proposed work aims at exploring a novel deep learning model for automated feature extraction from EEG signals to reduce the time, effort and complexity. From the literature study, it has been found that several ML techniques have been applied to extract features from EEG automatically but a proper comparative study on traditional and automatic feature extraction methods have not been put forward. In this paper, a deep learning model, convolutional neural network autoencoder (CNN-AE) is proposed for automatic feature extraction. These automated features are evaluated with several classification algorithms and compared with manual feature extraction technique for comparative analysis and feature optimisation.

The rest of the paper has been organised as follows – the background of the research domain and several related works, are described in Sect. 2. Section 3 contains detailed description of the experimental setup, data collection, analysis, feature extraction and classification techniques. Results along with the discussions are provided in Sects. 4 and 5 respectively. In the conclusion, limitations and future of this work are discussed in Sect. 6.

2   Background and Related Work

Literature indicates MWL as an important aspect of assessing human performance [12], whereas driving is a complex task performed by humans associated with several subsidiary tasks. Assessment of drivers' performance by quantifying MWL has been being performed for decades. There have been several means for measuring mental workload, but physiological measures are chosen often due to cheap and smaller technologies [13]. Physiological measures include respiration, electrocardiac activities, skin conductance, blood pressure, ocular measures, brain measures etc. Recently, Charles and Nixon stated that, brain measures in the form of EEG has been used for measuring MWL in most of the research works [12, 14]. Moreover, several studies have proven a strong correlation between MWL and EEG features both in time and frequency domain. Features like theta and alpha wave rhythms of EEG signal over the frontal and parietal sites respectively reflect significantly on the MWL variations [8, 15, 16].

Since the exploration of EEG signals, as a tool for measuring MWL, conventional techniques of feature extraction including statistical analysis and signal processing, have been in practice. Ahmed et al. proposed a non-linear approach of feature extraction using fractal dimensions to determine different brain conditions of participants [10]. In classifying motor imagery signals, Sherwani et al. used discrete wavelet transform analysis to extract feature from EEG signals [17] whereas Sakai used non-negative matrix factorisation [18]. Several techniques with time and frequency domain analysis have been proposed for feature extraction [19, 20]. Tzallas et al. proposed a method of extracting features from power spectrum density (PSD) of EEG segments by using Fourier transformation for epileptic seizure detection [11]. Individual alpha frequency (IAF) analysis has been adopted in several studies to adjust features of EEG signals [21]. Recently, Wen and Zhang proposed a genetic algorithm based feature search technique for multi-class epilepsy classification [22]. However, sufficient works have been presented on classifying MWL from EEG signal analysis where different ML algorithms were deployed after extracting features analytically. Use of several ML algorithms were found in the literature for classifying MWL such as Support vector machine (SVM) [23, 24], k-nearest neighbours (k-NN) [23], fuzzy-c means clustering [25], multi-layer perceptron (MLP) [23, 26], etc.

Extracting features automatically from EEG signals is a relatively new field of research. Researchers have deployed diverse range of deep learning (DL) algorithms, commonly termed as autoencoders (AE) to extract feature from EEG signals both with/without preprocessing. Recently, Wen et al. used deep convolutional neural network (CNN) for unsupervised feature learning from EEG signals after applying data normalisation for preprocessing. To assess the performance of their proposed model, several classification algorithms were used to classify epilepsy patients [27]. In several works, authors used stacked denoising autoencoder (SDAE) [28], long short-term memory (LSTM) [29] and deep belief network (DBN) [30] for feature extraction after applying PSD for preprocessing. Gou et al. extracted features by deployed genetic algorithm for classifying epilepsy with k-NN classifier. In this approach, discrete wavelet transformation

(DWT) was used for preprocessing of raw EEG signals [31]. In 2018, Shaha et al. investigated two different deep learning (DL) models, SDAE and LSTM, for extracting features from EEG signals without any preprocessing. Afterwards, MLP was used to classify cognitive load on the participants who were asked to perform learning task [23]. Ayata et al. [32] and Almogbel et al. [33], both the research groups used CNN autoencoder (CNN-AE) for extracting features from EEG signals for classifying arousal and MWL among participants.

Evidently, feature extraction from EEG signals using CNN-AE have been a popular technique among researchers for classification tasks from epilepsy and MWL domain. Moreover, several classification algorithms were further used to measure the effectiveness of the features extracted automatically. But, to our knowledge none of the works represented a comparative study about feature extraction through manual analysis and automatic extraction of features using DL techniques to compare the performance in workload classification particularly for driving situations.

3 Materials and Methods

3.1 Experimental Setup

The experiment was performed in a route going through urban areas at the periphery of Bologna, Italy. There were 20 participants in this experiment. All the participants were students of University of Bologna, Italy with mean age of 24 (±1.8) years and licensed for about 5.9 (±1) years on average. The participants were recruited for the study on voluntary basis. Only the male participants were selected to conduct a study with homogeneous experimental group. The experiment was conduct ed following the principles defined in the Declaration of Helsinki of 1975 (Revised in 2000). Informed consent and authorisation to use the recorded data was signed after proper description of the experiment was provided to the participants.

During the experiment a participant had to drive a car, Fiat 500L 1.3Mjt, with diesel engine and manual transmission, along the route illustrated in Fig. 1. In particular, the route consisted of three laps of a circuit about 2.5 km long to be covered with the daylight. The circuit was designed on the basis of evidences put forward in scientific literature [34,35]. In the designed circuit, there were two segments of interest in terms of road complexity and cognitive demand – (i) *Easy*, a straight secondary road serving residential area with an intersection halfway with the right-of-way; (ii) *Hard*, a major road with two roundabouts, three lanes, high traffic capacity and serving commercial area. This factor will be termed as "ROAD" in the following sections. Furthermore, a participant had to drive twice a day in the circuit, once during rush hour traffic and another in off-peak hour. This factor will be further termed as "HOUR" with two conditions *Normal* and *Rush*. This factor had been designed following the General Plan of Urban Traffic of Bologna, Italy. Table 1 refers the traffic flow intensity considered to design two experimental conditions in this study.

Fig. 1. The experimental circuit about 2.5 km long along Bologna roads. The red and yellow line along the route indicates 'Hard' and 'Easy' segments of the road respectively. The green arrow in the bottom-right corner shows the direction of driving from the starting and finishing point. (Color figure online)

Table 1. Traffic flow intensity in the experimental area during a day retrieved from General Plan of Urban Traffic of Bologna, Italy.

Transits	Total hour 14 h (6 ÷ 20)	Rush hour		Normal hour 12 h
		Morning (1230–1330)	Afternoon (1630–1730)	
Total	19385	2024	2066	15295
Frequency	–	2024	2066	12746

At the end of every experimental procedure consisting of a driving task of three laps twice during rush and normal hours, each participant was properly debriefed. The order of rush and normal hour condition had been randomised among the participants to avoid any order effect [36]. There were two segments in each lap, easy and hard referring to road complexity and task difficulties. During the whole experimental protocol physiological data in terms of brain activities through EEG has been recorded. A detailed description on recording of EEG signals has been given in the following sections. However, two very recent studies have been performed by Di Flumeri et al. following the same experimental procedure [15, 16].

3.2 Data Collection and Processing

EEG signals have been recorded using digital monitoring BEmicro systems provided by EBNeuro, Italy. Twelve EEG channels (FPz, AF3, AF4, F3, Fz, F4, P3, P7, Pz, P4, P8 and POz) were used to collect the EEG signals. The channels were placed on the scalp according to the 10–20 International System. The sampling frequency was 256 Hz for recording EEG signals. All the electrodes were referenced to both the earlobes and grounded to Cz site. Impedance was kept below 20 kΩ. During the experiment no signal conditioning were done, all the EEG signal processing were done offline. Events were recorded along with EEG signals to associate specific signals to different road and hour conditions.

Raw EEG signals were cropped referencing the events recorded; three laps for both *Normal & Rush* hours including *Easy & Hard* conditions. Furthermore, two ROAD-HOUR driving situations; *Easy-Normal* and *Hard-Rush* were selected for the classification of MWL since literature suggests that these conditions demand low and high MWL respectively [15]. Data of all the laps driven by the participants in the *Easy-Normal* and the *Hard-Rush* conditions were used for further analysis. EEG signals were sliced into 2 s (epoch length) segments by sliding window technique with a stride of 0.125 s keeping an overlap of 0.825 s between two continuous epoch. The windowing technique was performed to obtain a higher number of observations in comparison with the number of variable and respecting the condition of stationarity of the EEG signals [37]. Specific procedures of EEGLAB toolbox [38] have been used for slicing the recorded EEG signals. To remove different artefacts i.e. ocular and muscle movements etc. from the raw EEG signals ARTE algorithm by Barua et al. [39] has been used.

3.3   Feature Extraction

Two different types of feature extraction techniques i.e., manual and automatic were investigated in this study. In both the methods artefact handled EEG signals have been used. Firstly, the technique following traditional practices with filtering and signal processing methods has been used. Here, 25 relevant features were retrieved. Further, for the other approach DL was used to extract features from EEG signals. Here, 284 features were primarily extracted by CNN-AE. After analysing the feature importance based on random forest (RF) classifier 124 features were used for further tasks. Table 2 demonstrates the number of relevant features extracted from different techniques followed by description of two different feature extraction techniques.

Table 2. Number of features selected from different techniques.

Feature extraction	Number of features
Traditional	25
Deep learning	124

Traditional Approach. The process of feature extraction performed in this work is mostly motivated by the work done by Di Flumeri et al. [15]. Firstly, PSD has been calculated for each channel of each windowed epoch of ARTE cleaned EEG signals mentioned in Sect. 3.2. To calculate the PSD from the EEG signals, Welch's method [40] with Blackman-Harris window function was used on the same length of the epochs (2 s, 0.5 Hz frequency resolution). In particular, only the theta band (5–8 Hz) over the EEG frontal channels and the alpha band (8–11 Hz) over the EEG parietal channels, were considered as variables for the mental workload evaluation [8]. Then, to define EEG frequency bands of interest, IAF values were estimated with the algorithm developed by Corcoran et al. [21]. Figure 2 illustrates the final feature vectors generation, for each of the observations following the aforementioned sequence of steps.

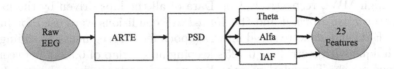

Fig. 2. Steps in traditional feature extraction technique.

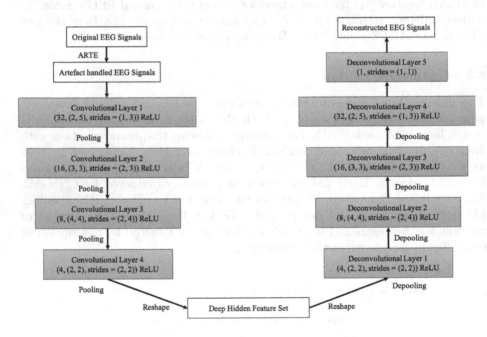

Fig. 3. Network architecture of the CNN-AE for feature extraction.

Deep Learning Approach. The CNN-AE architecture used for automatic feature extraction is shown in Fig. 3. The whole network is divided into two parts, (i) encoder and (ii) decoder. Encoder is comprised of a number of convolutional layer associated with pooling layers, finds deep hidden features from original signal. On the other hand, Decoder uses several deconvolutional layer to reconstruct the signal from the features. To assess the performance of the encoders, the quality of reconstructed signal from decoder is used. On the basis of this compressing and reconstructing, the whole model is trained. The developed encoder in this study, consists of four convolutional layers and four max-pooling layers. The decoder is designed in inverse order of the encoder. It contains five convolutional layers and four upsampling layers facilitating the depooling. Zero padding, batch normalisation and ReLU activation function have been used in each of the layers. The developed CNN-AE utilised RMSprop optimisation with a learning rate of 0.002 and binary cross-entropy as the loss function. After a successful learning procedure, CNN-AE extracted 284 features from the experimental EEG signals.

3.4   Classification of MWL

After extracting features from two different methods, several classifiers were deployed to classify MWL. Table 3 provides the list of classifiers and the values of their prominent parameters.

Table 3. Parameters used in different classifiers.

Classifier	Parameter details
Support Vector Machine (SVM)	$kernel = \text{'}rbf\text{'}$
k-Nearest Neighbout (kNN)	$k = 5$
Random Forest (RF)	$max\_depth = 5, n\_estimators = 100$
Multi-layer Perceptron (MLP)	$hidden\_layer = 100, activation = relu$

Before classifying MWL, to reduce the dimension of the feature set further, feature importance was calculated using RF classifier. Different number of features were selected from 284 features depending on different threshold values and deployed for classifying MWL with SVM classifier on the training data set. It was observed that there was variation in accuracy. Finally, by imposing 0.003 threshold on feature importance 124 relevant features were finalised that reduced the feature set by more than half but increased accuracy. For the both the classifiers, parameters given in Table 3 were used. Figure 4 illustrates the change of accuracy for different threshold values of feature importance to select features for classification.

4   Result and Evaluation

All the observations with relevant features from the EEG signals were divided into training and testing set considering 80% and 20% of the data respectively. The training set was used to train the model and the testing set was used to validate the accuracy of MWL classification. Several common classifiers stated Table 3 were deployed to verify the effectiveness of the features obtained by traditional method and CNN-AE. For measuring classification performance, average overall accuracy, balanced classification rate (BCR) or balanced accuracy and

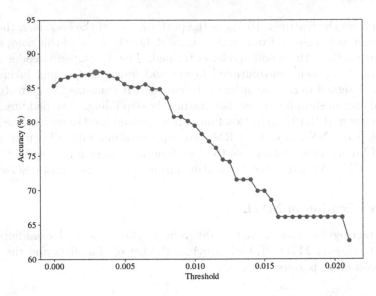

Fig. 4. Variation in classification accuracy with respect to the change of threshold on feature importance values. Highest average accuracy 87.30% was found for 0.003 (point marked with red dot) as threshold on feature selection. (Color figure online)

F_1 score were calculated for each of classifiers and features extracted by different methods. Tables 4 and 5 contains the values for performance measures of classification from traditionally extracted features and CNN-AE extracted features respectively. It has been observed that features extracted from CNN-AE produced better performance measures for all the classifiers. In particular, SVM classified MWL with the highest overall accuracy of 87%.

Table 4. Average performance measures of classifiers applied on traditionally extracted features.

Classifiers	Accuracy	BCR	F_1 score
SVM	0.5388	0.5388	0.5146
kNN	0.6420	0.6420	0.6486
RF	0.6414	0.6414	0.6442
MLP	0.7083	0.7083	0.7151

Table 5. Average performance measures of classifiers applied on features extracted by CNN-AE.

Classifiers	Accuracy	BCR	F_1 score
SVM	0.8700	0.8700	0.8730
kNN	0.7737	0.7737	0.7912
RF	0.8049	0.8049	0.8197
MLP	0.8504	0.8504	0.8527

To investigate the performance of the classifiers further, Specificity and Sensitivity were calculated and illustrated in Fig. 5. It has been clearly visible from the figure that both the scores for CNN-AE features were higher than traditionally extracted features.

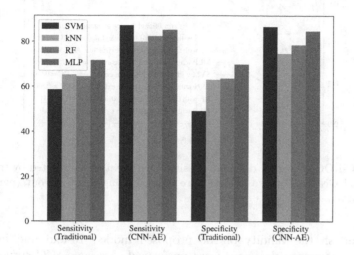

Fig. 5. MWL classification results in terms of Sensitivity and Specificity.

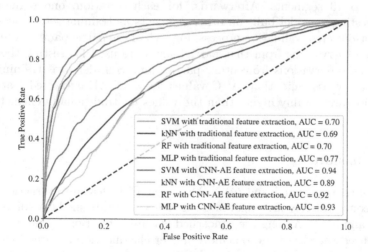

Fig. 6. AUC-ROC curves for different classifiers with features extracted by traditional methods and CNN-AE where models were trained using 10-fold cross-validation.

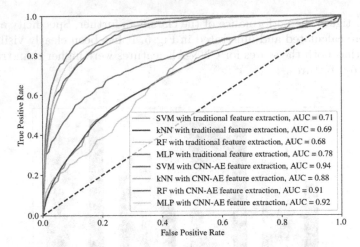

Fig. 7. AUC-ROC curves for different classifiers with features extracted by traditional methods and CNN-AE where models were trained using leave-one-out (participant) cross-validation.

To establish the validity of the proposed model, 10-fold and leave-one-participant-out cross validations were performed. Average AUC curves on the cross validations are illustrated in Figs. 6 and 7 where SVM classifier has the highest AUC in both. For 10-fold cross validation, all the observations were divided into 10 segments. Afterwards, for each iteration, one segment was used for testing a model built on other segments as training set. In leave-one-participant-out cross validation process, for each of the participants of the experiment, the observations from that participant were used for testing the model build on the observations from other participants considered as training data. For both the cross validation, AUC values for CNN-AE extracted features in classification are notably higher than the values for traditionally extracted feature.

5   Discussion

In this study, traditional and CNN-AE based EEG feature extraction methods were comparatively investigated using four well established classifiers; SVM, kNN, RF and MLP. Among the concerned feature extraction techniques, CNN-AE influenced the classifiers to achieve higher classification accuracy and other performance measures. Initially, the number of features extracted from CNN-AE were substantially higher than the features extracted through traditional methods but with feature selection mechanism, the feature set was approximately reduced to half resulting improvement in the accuracy measures of all classifiers. From different performance measures demonstrated in Sect. 4, it has been shown that SVM achieves higher accuracy in classifying MWL from EEG signals irrespective of feature extraction technique.

In case of classifier models for MWL classification used in related works, many factors affect the performance of the model. Generally, if there remains a clear correlation between characteristics of data and class labels, the deployed classifier achieves higher accuracy in prediction. But, in case of MWL classification for drivers' while driving in real life or simulator, the probability of noise being recorded with the EEG signals is quite high due to eye movement, power signals, miscellaneous interference etc. In practice, the noises are termed as artefacts. In traditional feature extraction methods, removing these artefacts from data along with different inter- and intra-individual variability require huge manual effort and processing. According to the characteristics of deep learning, its layer can find out hidden features laid in a data responsible of assigned labels. Here, from the results of this study it can be established that, CNN-AE or any deep learning mechanism can produce feature set from EEG signals, that would be equivalent or better than the feature set extracted manually with less effort keeping aside the preprocessing and artefact handling tasks. Primarily, the proposed CNN-AE produced an extensive set of features. An intuitive investigation on the feature selection with RF Classifier and imposing threshold on feature importance produced considerably shorter feature vector with higher classification accuracy. Further investigation on feature selection in this domain can produce more robust set of relevant features.

The recorded data from experimental protocol was balanced in terms of class labels. Each of the participants attempted driving for different ROAD and HOUR condition once. The recorded EEG signals formed the initial labelled balanced data. For further investigation, the raw EEG signals were segmented into overlapping epochs to increase the amount of observations keeping the core characteristics of the data. This operation facilitated this data-driven study by increasing the amount data with a trade-off for balanced data. Due the uneven driving duration among the participants, the number of windowed epochs varied from participant to participant as well as for different study factors resulting the data as an imbalanced data. Performance measures illustrated in Sect. 4 were chosen from prescribed measures for imbalanced data by Tharwat [41].

6   Conclusion

This paper presents a new hybrid approach for automatic feature extraction from the EEG signals and demonstrated with MWL classification. The main contribution of this paper can be represented in three folds: (i) CNN method is used to extract features automatically from artefact handled EEG signals, (ii) RF is used for feature selection and (iii) several machine learning algorithms are used to classify drivers' mental workload on CNN based feature sets. This new hybrid approach is compared with traditional feature extraction approach considering four machine learning classifiers, i.e. SVM, kNN, RF and MLP. According to the outcome of the both 10-fold and leave-one-participant-out cross validation, SVM outperforms other classifiers with CNN-AE extracted features. One advantage of CNN-AE for feature extraction is that it works directly on the artefact handled data sets i.e. additional signal processing, individual feature extraction etc. are not needed, thus reducing time in manual work. More experimental work with large and heterogeneous data set is planned for future work to increase the performance of the proposed method and extract features directly from raw EEG signals. Moreover, classifying MWL in real time using the proposed approach and suggesting external actions to mitigate road casualty is the final goal of the planned research works.

Acknowledgement. This article is based on work performed in the project Brain-SafeDrive. The authors would like to acknowledge the Vetenskapsrådet (The Swedish Research Council) for supporting the BrainSafeDrive project. The authors are also very thankful to Prof. Fabio Babiloni of BrainSigns. They would also like to acknowledge the extraordinary support of Dr. Gianluca Borghini & Dr. Pietro Aricó in experimental design & data collection. Further, authors would like to acknowledge the project students, Casper Adlerteg, Dalibor Colic & Joel Öhrling for their contribution to test the concept.

References

1. Kar, S., Bhagat, M., Routray, A.: EEG signal analysis for the assessment and quantification of driver's fatigue. Transp. Res. Part F Traffic Psychol. Behav. **13**(5), 297–306 (2010)
2. Thomas, P., Morris, A., Talbot, R., Fagerlind, H.: Identifying the causes of road crashes in Europe. Ann. Adv. Automot. Med. **57**, 13–22 (2013)
3. Kim, H., Yoon, D., Lee, S.J., Kim, W., Park, C.H.: A study on the cognitive workload characteristics according to the ariving behavior in the urban road. In: 2018 International Conference on Electronics, Information, and Communication (ICEIC), pp. 1–4. IEEE (2018)
4. Brookhuis, K.A., de Waard, D.: Monitoring drivers' mental workload in driving simulators using physiological measures. Accid. Anal. Prev. **42**(3), 898–903 (2010)
5. Almahasneh, H., Kamel, N., Walter, N., Malik, A.S.: EEG-based brain functional connectivity during distracted driving. In: 2015 IEEE International Conference on Signal and Image Processing Applications (ICSIPA), pp. 274–277. IEEE (2015)
6. Moustafa, K., Luz, S., Longo, L.: Assessment of mental workload: a comparison of machine learning methods and subjective assessment techniques. In: Longo, L., Leva, M.C. (eds.) H-WORKLOAD 2017. CCIS, vol. 726, pp. 30–50. Springer, Cham (2017). https://doi.org/10.1007/978-3-319-61061-0_3
7. Aricò, P., Borghini, G., Di Flumeri, G., Colosimo, A., Pozzi, S., Babiloni, F.: A passive brain-computer interface application for the mental workload assessment on professional air traffic controllers during realistic air traffic control tasks. In: Progress in Brain Research, vol. 228, pp. 295–328. Elsevier (2016)
8. Aricò, P., Borghini, G., Di Flumeri, G., Sciaraffa, N., Colosimo, A., Babiloni, F.: Passive BCI in operational environments: insights, recent advances, and future trends. IEEE Trans. Biomed. Eng. **64**(7), 1431–1436 (2017)
9. Begum, S., Barua, S.: EEG sensor based classification for assessing psychological stress. Stud. Health Technol. Inform. **189**, 83–88 (2013)
10. Ahmad, R.F., et al.: Discriminating the different human brain states with EEG signals using fractal dimension- a nonlinear approach. In: 2014 IEEE International

Conference on Smart Instrumentation, Measurement and Applications (ICSIMA), pp. 1–5. IEEE (2014)

11. Tzallas, A.T., Tsipouras, M.G., Fotiadis, D.I.: Epileptic seizure detection in eegs using time-frequency analysis. IEEE Trans. Inf. Technol. Biomed. **13**(5), 703–710 (2009)

12. Charles, R.L., Nixon, J.: Measuring mental workload using physiological measures: a systematic review. Appl. Ergon. **74**, 221–232 (2019)

13. Guzik, P., Malik, M.: ECG by mobile technologies. J. Electrocardiol. **49**(6), 894–901 (2016)

14. Barua, S., Ahmed, M.U., Begum, S.: Classifying drivers' cognitive load using EEG signals. In: pHealth, pp. 99–106 (2017)

15. Di Flumeri, G., et al.: EEG-based mental workload neurometric to evaluate the impact of different traffic and road conditions in real driving settings. Front. Hum. Neurosci. **12**, 509 (2018)

16. Di Flumeri, G., et al.: EEG-based mental workload assessment during real driving: a taxonomic tool for neuroergonomics in highly automated environments. In: Neuroergonomics, pp. 121–126. Elsevier (2019)

17. Sherwani, F., Shanta, S., Ibrahim, B., Huq, M.S.: Wavelet based feature extraction for classification of motor imagery signals. In: 2016 IEEE EMBS Conference on Biomedical Engineering and Sciences (IECBES), pp. 360–364. IEEE (2016)

18. Sakai, M.: Kernel nonnegative matrix factorization with constraint increasing the discriminability of two classes for the EEG feature extraction. In: 2013 International Conference on Signal-Image Technology & Internet-Based Systems, pp. 966–970. IEEE (2013)

19. Barua, S.: Multivariate Data Analytics to Identify Driver's Sleepiness, Cognitive load, and Stress. Ph.D. thesis, Mälardalen University (2019)

20. Begum, S., Barua, S., Ahmed, M.U.: In-vehicle stress monitoring based on EEG signal. Int. J. Eng. Res. Appl. **7**(7), 55–71 (2017)

21. Corcoran, A.W., Alday, P.M., Schlesewsky, M., Bornkessel-Schlesewsky, I.: Toward a reliable, automated method of individual alpha frequency (IAF) quantification. Psychophysiology **55**(7), e13064 (2018)

22. Wen, T., Zhang, Z.: Effective and extensible feature extraction method using genetic algorithm-based frequency-domain feature search for epileptic EEG multiclassification. Medicine **96**(19), 1–11 (2017). https://doi.org/10.1097/MD.0000000000006879

23. Saha, A., Minz, V., Bonela, S., Sreeja, S.R., Chowdhury, R., Samanta, D.: Classification of EEG signals for cognitive load estimation using deep learning architectures. In: Tiwary, U.S. (ed.) IHCI 2018. LNCS, vol. 11278, pp. 59–68. Springer, Cham (2018). https://doi.org/10.1007/978-3-030-04021-5_6

24. Zarjam, P., Epps, J., Lovell, N.H.: Beyond subjective self-rating: EEG signal classification of cognitive workload. IEEE Trans. Auton. Ment. Dev. **7**(4), 301–310 (2015)

25. Das, D., Chatterjee, D., Sinha, A.: Unsupervised approach for measurement of cognitive load using EEG signals. In: 13th IEEE International Conference on BioInformatics and BioEngineering, pp. 1–6. IEEE (2013)

26. Zarjam, P., Epps, J., Chen, F.: Spectral EEG features for evaluating cognitive load. In: 2011 Annual International Conference of the IEEE Engineering in Medicine and Biology Society, pp. 3841–3844. IEEE (2011)

27. Wen, T., Zhang, Z.: Deep convolution neural network and autoencoders-based unsupervised feature learning of EEG signals. IEEE Access **6**, 25399–25410 (2018)

28. Yin, Z., Zhang, J.: Recognition of cognitive task load levels using single channel EEG and stacked denoising autoencoder. In: 2016 35th Chinese Control Conference (CCC), pp. 3907–3912. IEEE (2016)
29. Manawadu, U.E., Kawano, T., Murata, S., Kamezaki, M., Muramatsu, J., Sugano, S.: Multiclass classification of driver perceived workload using long short-term memory based recurrent neural network. In: 2018 IEEE Intelligent Vehicles Symposium (IV), pp. 1–6. IEEE (2018)
30. Xiang, L., Zhang, P., Song, D., Yu, G., et al.: EEG based emotion identification using unsupervised deep feature learning. In: SIGIR2015 Workshop on Neuro-Physiological Methods in IR Research, 13 August 2015 (2015)
31. Guo, L., Rivero, D., Dorado, J., Munteanu, C.R., Pazos, A.: Automatic feature extraction using genetic programming: an application to epileptic EEG classification. Expert. Syst. Appl. **38**(8), 10425–10436 (2011)
32. Ayata, D., Yaslan, Y., Kamasak, M.: Multi channel brain EEG signals based emotional arousal classification with unsupervised feature learning using autoencoders. In: 2017 25th Signal Processing and Communications Applications Conference (SIU), pp. 1–4. IEEE (2017)
33. Almogbel, M.A., Dang, A.H., Kameyama, W.: EEG-signals based cognitive workload detection of vehicle driver using deep learning. In: 2018 20th International Conference on Advanced Communication Technology (ICACT), pp. 256–259. IEEE (2018)
34. Paxion, J., Galy, E., Berthelon, C.: Mental workload and driving. Front. Psychol. **5**, 1344 (2014)
35. Verwey, W.B.: On-line driver workload estimation. Effects of road situation and age on secondary task measures. Ergonomics **43**(2), 187–209 (2000)
36. Kirk, R.E.: Experimental Design. Handbook of Psychology, 2nd ed. (2012)
37. Elul, R.: Gaussian behavior of the electroencephalogram: changes during performance of mental task. Science **164**(3877), 328–331 (1969)
38. Delorme, A., Makeig, S.: EEGLAB: an open source toolbox for analysis of single-trial EEG dynamics including independent component analysis. J. Neurosci. Methods **134**(1), 9–21 (2004)
39. Barua, S., Ahmed, M.U., Ahlstrom, C., Begum, S., Funk, P.: Automated EEG artifact handling with application in driver monitoring. IEEE J. Biomed. Health Inform. **22**(5), 1350–1361 (2017)
40. Solomon Jr., O.: PSD computations using welch's method. NASA STI/Recon Technical Report N **92** (1991)
41. Tharwat, A.: Classification assessment methods. Appl. Comput. Inform. (2018, in press). https://doi.org/10.1016/j.aci.2018.08.003

Hybrid Models of Performance Using Mental Workload and Usability Features via Supervised Machine Learning

Bujar Raufi[✉] ⓘ

South East European University, Tetovo 1200, North Macedonia
b.raufi@seeu.edu.mk

Abstract. Mental Workload (MWL) represents a key concept in human performance. It is a complex construct that can be viewed from multiple perspectives and affected by various factors that are quantified by different collection of methods. In this direction, several approaches exist that aggregate these factors towards building a unique workload index that best acts as a proxy to human performance. Such an index can be used to detect cases of mental overload and underload in human interaction with a system. Unfortunately, limited work has been done to automatically classify such conditions using data mining techniques. The aim of this paper is to explore and evaluate several data mining techniques for classifying mental overload and underload by combining factors from three subjective measurement instruments: System Usability Scale (SUS), Nasa Task Load Index (NASATLX) and Workload Profile (WP). The analysis focused around nine supervised machine learning classification algorithms aimed at inducing model of performance from data. These models underwent through rigorous phases of evaluation such as: classifier accuracy (CA), receiver operating characteristics (ROC) and predictive power using cost/benefit analysis. The findings suggest that Bayesian and tree-based models are the most suitable for classifying mental overload/underload even with unbalanced data.

Keywords: Usability · Hybrid models of performance · Supervised machine learning · Nasa Task Load Index · Workload Profile · System Usability Scale

1 Introduction

Mental Workload (MWL) represents a fundamental concept in human factors. It is a complex construct that can be viewed from different perspectives including measurements, prediction and consequences; which can be affected by several factors measurable with various methods [1–3]. Different approaches have been followed to aggregate these factors into defining a unique index of MWL. However, several obstacles appear when we try to understand, determine and describe which factors contribute to system usability and MWL towards building a robust model that have general applicability for predicting user performance [4]. State-of-the-art computational models are rather ad hoc, and their application is limited and domain specific [5–7, 31, 34]. Furthermore, the vast majority of models are mainly theory-driven, meaning that

© Springer Nature Switzerland AG 2019
L. Longo and M. C. Leva (Eds.): H-WORKLOAD 2019, CCIS 1107, pp. 136–155, 2019.
https://doi.org/10.1007/978-3-030-32423-0_9

an inference and generalizations are driven from a set of measurable factors, theoretically related to MWL, and aggregated via hand crafted rules or techniques. Eventually, this inference is theoretically related to human performance [8]. It is worth noting that limited work has been done towards the development of data-driven models of MWL. This means that computational models of mental workload that are induced directly from data and human performance derived from it are lacking. The reason is justified by the fact that MWL is still an inadequately defined construct, supporting the application of deductive research methods. Another reason is that MWL is a five decades old paradigm, which in its core, machine learning methodologies and algorithms were not given a serious attention [8, 30]. Only in the last two decades, with the proliferation of Machine Learning (ML) application in different disciplines, researchers initiated the exploration of MWL using data-driven or inductive research methodologies [9–13].

This paper attempts to model system usability and MWL by employing machine learning algorithms. It makes use of a dataset, which comprises a holistic approach related to MWL and system usability, employing features coming from popular self-rating instruments like the NASA Task Load Index (NASATLX), the Workload Profile (WP) and the Subjective Usability Scale (SUS). The rest of this paper is organized as follows: Sect. 2 presents related work concerning the use of Machine Learning approaches for usability and MWL, while Sect. 3 presents the design of an experiment following the popular data mining CRISP-DM methodology and Sect. 4 concludes the paper.

2 Related Work

Mental workload (MWL) is a fundamental design concept in Human-Computer Interaction (HCI), Ergonomics (Human Factors) and in Cognitive Psychology. MWL represents an intrinsically complex and multilevel concept and as such, there is no unified and widely accepted definition. However, it can be defined as the quantification of mental cost of performing tasks in order to predict operator and system performance under a finite time frame [4, 14, 15]. Various methods have been proposed for directly quantifying this construct. These methods can be clustered into three groups [8]. *Subjective measures*: which relies on the analysis of the subjective feedback information provided by humans interacting with an underlying task and system. The feedback is usually in the form of a post-task survey or questionnaire. The most well-known subjective measurement approaches are the NASA Task Load Index (NASATLX) [15], the Workload profile (WP) [14], and the Subjective Workload Assessment Technique (SWAT) [14]; *Task performance measures*, often referred to as primary and secondary tasks measures. It focuses on the objective performance measurement of a human related to an underlying task. Time completion of a task, the reaction time to secondary tasks and the number of errors on the primary task are examples of such measures, together with tracking and analyzing different actions performed by a user during a primary task. *Physiological measures* that are based upon the analysis of physiological responses of the human body. Examples include EEG (electroencephalogram), MEG (magnetoencephalogram), Brain Metabolism, Endogenous Eye blinks, Pupil diameter, heart rate measures or electrodermal responses [16]. From the conducted literature review concerning usability, MWL and machine learning techniques, we have identified two research paths. The first one focus on psychological measurements on classification of mental tasks from EEG signals [23, 24, 32, 33] Some attempts in this "path of attack" focus on more optimized approach on assessing temporal variations of mental workload by implying feature reduction of psychological human responses using SVM based clustering and classification techniques [25], identifying workload conditions using various machine learning approaches such as adaptive support vector machines (SVMs) [26] or ensemble learning methods [27]. The second research path follows the approach of mental workload modeling with data-driven inductive research methodologies [22, 28]. Some approaches in this line even tries to bring MWL to HCI by indirectly classifying MWL through affect detection in interfaces by empirically evaluating four machine learning methods like Nearest Neighbor, Regression Tree (RT), Bayesian Network and Support Vector Machine (SVM) [29]. It is worth mentioning that, most of the research above fail to address the wholistic approach on measuring user performance by combining usability and mental workload (MWL) features. Our hybrid approach on modeling use performance in web interfaces is an attempt in this direction.

3 Machine Learning Assessment of User Performance

Assessing user performance with system usability and Mental Workload (MWL) through machine learning techniques requires a well-founded data mining methodology that should be pursued. For this purpose, we utilized the CRISP-DM (**CR**oss **I**ndustry **S**tandard **P**rocess for **D**ata **M**ining) methodology [21]. The main goal of this work is to apply machine learning techniques for semi-automated user performance classification in Web interfaces. Scientific experiments and results indicate that mental underload and overload negatively affect human performance [22]. Consequently, creating a training dataset for building a classification model capable of efficiently detecting certain performance conditions such as mental overload/underload is important. This paper also focuses on classification model evaluation in order to select the most robust model for such task. For this purpose, we utilized user performance estimation using mental workload and usability features in web interfaces comprising the following self-reporting assessment instruments:

- The System Usability Scale (SUS)
- The Nasa Task Load Index (NASATLX) and
- The Workload Profile (WP) [19] derived from Multiple Resource Theory [17, 18]

The *System Usability Scale (SUS)* represent a simple ten elements measurement scale for characterizing a global view on subjective usability assessment. The SUS serves as a Likert scale based on "forced" questions bound to an agreement or disagreement levels. This means that initially a statement is made, and latter, user is forced to respond with an answer bound to a 5- or 7-point scale and as such, it is a "quick and dirty" tool for measuring usability [20]. In the dataset subject to our analysis [22], the SUS values are represented in an interval of a range $[0, \ldots 100] \in \Re$ given formally as:

$$SUS: [0, \ldots 1] \in \Re i_1 = \{1, 3, 5, 7, 9\} i_2 = \{2, 4, 6, 8, 10\}$$

$$SUS = \frac{1}{10} \cdot \left[\sum_{i_1} (SUS_i) + \sum_{i_2} (100 - SUS_i) \right]$$

Individual scores for SUS are calculated separately for even and odd questions. For odd questions, the score contribution is the SUS scale minus one, and for even questions the contribution is five minus the SUS scale. The questionnaire used for assessing the SUS as used in [22] is reintroduced in Table 6 in the appendix. *The NASA Task Load Index (NASATLX)* is a self-reporting subjective measurement instrument consisted of six factors believed to influence the mental workload: mental, physical and temporal demand, effort, performance and stress [15]. It is coupled with a pairwise comparison for assessing the strength of each of the above 6 factors. The final NASATLX index is calculated as an average of each factors f_i together with their respective weights w_i:

$$NASATLX: [0, \ldots 1] \in \Re NASATLX = \frac{1}{15} \left(\sum_{i=1}^{6} f_i \times w_i \right)$$

The number 15 appearing in the equation corresponds to the number of answers set where each factor from NASATLX was selected. Questions used for generating NASATLX index is given in appendix's Table 7.

Workload Profile (WP) is based on multiple resource theory [17, 18] where individuals are seen as having capacities such as: level of information processing (perceptual/central and response selection and execution), code of information processing (spatial/verbal) and inputs (visual/auditory processing) and output (manual and speech output). The Workload Profile (WP) quantification in the dataset we analyzed is an aggregation between all the above-mentioned rates given as:

$$WP: [0, \ldots 8] \in \Re WP = \sum_{i=1}^{8} f_i$$

Table 8 in the appendix depicts the questionnaire used for Workload Profile (WP). The experiment for collecting data from the above-mentioned questionnaires used a sample of 40 participants divided into two groups of 20 each, gender-balanced in female and male users and ages between 20 to 35 years. Participants were asked to execute a set of 9 web-based tasks related to information seeking as naturally as they could over the period of 2 or 3 sessions of approximately 45/70 min each. The sessions were scheduled on different non-consecutive days. The complete list of tasks is given in more details in [22]. The experimental procedure was designed in such a way to investigate how the perception of usability between the two groups interacts with subjective assessment of mental workload of users as well as the system usability. While the procedure was ongoing, users could not interact with instructors during the tasks. The final dataset resulted in containing 405 instances and 66 attributes which have been selected from experiments related to NASATLX, SUS and WP instruments elaborated extensively in [22]. Table 1 depicts the main descriptive for the important variables describing the three instruments from the sense of data variance, kurtosis and skewness.

Table 1. Default variance, standard deviation and skewness for SUS, WP and NASATLX

Variable	N	Mn. (μ)	Std (σ)	Var. (s)	Kurtosis	S.E. Kurt	Skewness	S.E. Skew	Min	Max
WP	405	37.53	15.03	225.98	−0.35	0.24	−0.16	0.12	0	72.75
SUS	405	74.23	19.29	372.13	0.32	0.24	−0.91	0.12	11	100
NASATLX	405	46.56	19.36	374.99	−0.23	0.24	0.02	0.12	0.8	99.93
Avg_Perf	405	63.91	17.83	318.04	−0.09	0.24	−0.81	0.12	15.81	93.71

Table 2 also details the internal consistency of the data measured with Cronbach's alpha value. From the dataset we have a moderate data reliability of 59% considering that we have three different instruments for measuring user performance. The reason lies in the fact that the dataset has five independent feature sets having different data characteristics (mean, standard deviation and median), which are illustrated in Fig. 1.

Table 2. Data reliability of the analyzed dataset

		N	[%]
# Instances	Valid	288	71.11
	Excluded	117	28.89
	Total	405	100
Cronbach's Alpha		0.59	59
# Attributes		67	

The data involved in our analysis comprises from the MWL measurements like NASATLX and WP as well as SUS usability data. For data exploration purposes, an Analytic Base Table (ABT) for both MWL and usability instruments was generated in order to discover the nature and characteristics of data at hand such as type of its features, data values and ranges. The dataset reveals 5 independent feature sets each of them having 5, 10, 5, 8 and 10 attributes respectively. The first feature set from data involves the basic NASATLX questions, the second feature set comprises the pairwise comparison of NASATLX data, the third has the total preferences of pairwise comparison (weighted NASTLX), the fourth feature set is the original workload profile (WP) data and the last data feature is the original system usability scale (SUS). Figure 1 illustrates the data characteristics of the four features from the perspective of mean, standard deviation and median.

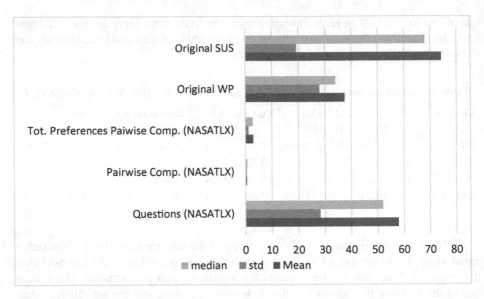

Fig. 1. Characteristics of the five independent features from data

The complete descriptive of data of all independent features are given in Table 9 of the appendix.

3.1   Data Preparation

In the data preparation phase, from the initial dataset, a new attribute called *PerfClass* for user performance classification with three nominal values *{underload, normal_load and overload}* is introduced. This attribute is based on the average performance values from the three subjective self-reporting usability and MWL instruments (SUS, NASATLX and WP). In order to quantify and separate the *PerfClass* nominal values, an attribute called *Avg_Perf* is introduced to grasp the overall user performance from both usability and MWL instruments. Figure 2 depicts average user performance in comparison to SUS, WP and NASATLX per each task. The total number of tasks given to users are 9 based on the experimentation done in [22].

The average performance values are in the range between $[0, \ldots, 100]$. When the performance value is below 25 of the average performance, we have an overload condition considering that user could not complete a particular task and when the given performance was above 75, the task is considered to be easy for completion resulting in mental underload [4]. The cutoff values of 25/75 are taken from the workload spectrum separation into four regions as elaborated in [4].

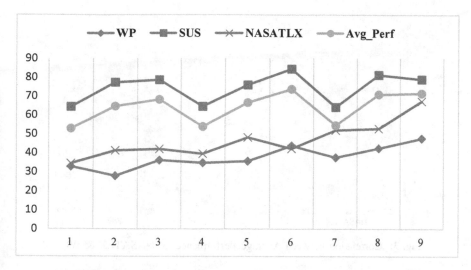

Fig. 2. Average user performance vs. SUS, WP and NASATLX instruments

3.2   Model Evaluation

In order to get better understanding the extent to which the hybrid model can explain the user performance, a concurrent validity between MWL subjective instruments (NASATLX & WP) and usability instruments (SUS) against Average Performance (*Avg_Perf*) has been used.

To measure the concurrent validity of the selected MWL instruments (NASATLX, WP), the Pearson's correlation coefficient has been selected as it evaluates the monotonic relationship between the two continuous MWL indexes against *PerfClass* attribute. Table 3 illustrates a relatively strong statistical significance between NASATLX - Avg_Perf and WP - Avg_Perf.

Table 3. Concurrent validity: correlation of NASATLX & WP vs. Average Performance

	NASATLX	WP
Average Performance (Avg_Perf)	0.74	0.72
Sig. (2-tailed)	0.0015	0.001

The correlation between NASATLX, WP and Avg_Perf is illustrated in Fig. 3. The blue and green data points on top right and bottom left corner represents performance underload and overload values.

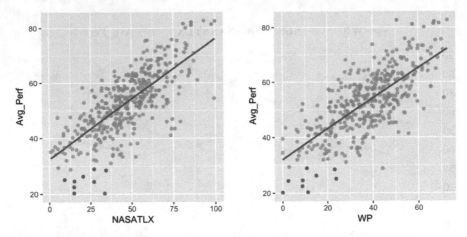

Fig. 3. Correlation between Average Performance vs. NASATLX & WP

The results on correlation between System Usability Scale (SUS) and average user performance is given in Table 4

Table 4. Concurrent validity: correlation of SUS vs. Average Performance

	SUS
Average Performance (Avg_Perf)	0.48
Sig. (2-tailed)	0.0001

Figure 4 illustrates a moderate correlation between SUS and average user performance.

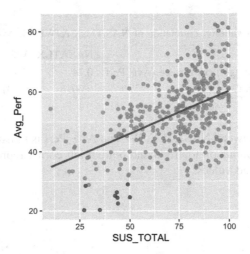

Fig. 4. Correlation between Average Performance vs. SUS

The aim is to select the most robust and accurate classifier that will exhibit the property to correctly classify user performance as underload or overload respectively. Considering only the classifier accuracy parameter is insufficient because of the fact that from the dataset, out of 405 instances, the *Avg_Perf* class has only 11 instances as Underload and 5 as Overload values. Figure 5 illustrates the histogram of distribution of all 405 instances across the nominal values of the *PerfClass* attribute.

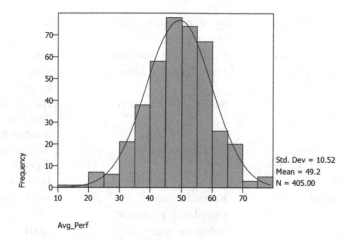

Fig. 5. Distribution of instances across the 3 groups of user performance from PerfClass

From the figure above, it can be seen that performance overload and underload are on the edges of the normal distribution, which entails further analysis to be done for selecting the best classifier for these two nominal values for the main classification class. In pursuance of selecting the best classifier that shows a high degree of accuracy during classification, we have entailed a Kappa statistics for measuring the inter rates agreement of categorical data in our dataset, Receiver Operating Characteristics (ROC) and Cost Benefit analysis for evaluating the robustness of machine learning techniques. The classifiers evaluated for the potential Machine Learning techniques suitable for classifying user performance are: Rule based classifiers (ZeroR and JRip), Bayesian Models (Naïve Bayes, Bayes Nets), Lazy Classifiers (k-Nearest Neighbor KNN), Support Vector Machine Models (SVM) and Tree Based Classification models (Decision Trees, Random Forest). Initial parameters set for all classifiers are summarized as in the Table 5 given below:

Table 5. Initial classifier parameters for evaluation

Learning model	Classifier	Parameter type	Value
Rule Based	ZeroR	Decimal places for the predicted numeric mean	2
	JRip	Folds number.	10
		Total weight of an instance in a rule	2.0
		No. of optimization runs	2
		Use of pruning	True
Bayesian Models	Naïve Bayes	Use of kernel estimator	True
	Bayes Nets	Use of kernel estimators	Simple and Bayes Kernel
		Search algorithm used	Hill climbing Simulated Annealing
Lazy Models	K-Nearest Neighbor (KNN)	Distance weighting	No distance, 1-distance weights
		NN Search Algorithm	Linear NN Search
Function Based	SVM	Kernel function	PolyKernel
		Complexity parameter	1.0
		Tolerance parameter	0.001
	ANNs	Hidden layer	50
		Learning rate	0.3
		Momentum	0.2
Tree Based	Decision Trees	Confidence pruning factor Info Gain/Gini Index Split,	0.25
		maxDepth:	5
		Minimum split:	2
		minBucket:	1
	Random Forest	Attribute Importance	True
		Number of randomly chose features	3

Because of the small dataset, a holdout method with 10-fold cross validation has been implied for all classifiers in order to properly determine the accuracy. The 10-fold CV generates 10 surrogate models out of which return the average model from these ten as the final one.

3.3   Evaluation

Evaluating the classifier accuracies from the standpoint of correlation significance through Kappa Statistic, Precision, recall and f-score rates is depicted in Fig. 6.

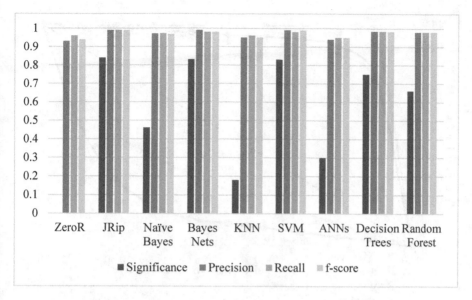

Fig. 6. Classifier accuracy for different machine learning algorithms

The results from the overhead figure indicate that almost all the classification methods show an outstanding classification accuracy for *normal_load* value of *Perf-Class* on the dataset, however it seriously lacks ability to detect performance overload and underload which is the goal in our approach. It is obvious that some trivial models are not suitable and efficient to be applied for discovering overload and underload values of the class. This fact clearly indicates that the accuracy cannot be applied for assessing the usefulness of classification models that are built using unbalanced datasets. For this purpose, a good approach is to use the significance of agreement of the nominal values in the main class through "Kappa statistic", which is zero for the case of ZeroR. Kappa statistic is an analog to correlation coefficient. Its value is zero for the lack of any relation in comparison to one for very strong statistical relation between the class label and attributes of instances, i.e. between the class *PerfClass* {*normal_load, overload, underload*} and the values of their descriptors. For this purpose, from the figure there are models with approximately the same classification accuracy, however they show great significance through Kappa statistic. Another useful statistical characteristic is using Receiver Operating Characteristic (ROC) analysis. The ROC analysis is used to evaluate the classifier performance by measuring the true Positive Rate (TPR) and False Positive Rates (FPR) received by the classifier. In detail, the ROC is considered, for which the value near 0. that a model behaves as a random model and values near FPR indicate poor classifier performance and values favoring TPR indicate good classifier performance.

Figure 4 depicts the Research Operations Curve (ROC) for all the learning methods for the classifiers that are object to analysis.

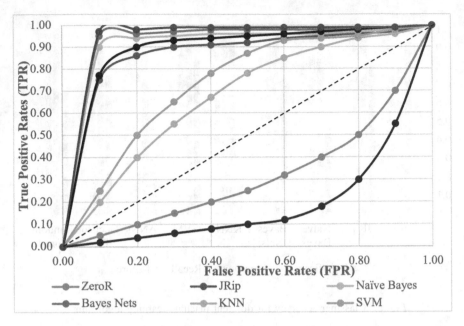

Fig. 7. ROC analysis for the classifiers.

Having the above-mentioned aspects in mind, a more thorough investigation is done in order to select more accurate and robust classification model able to detect performance overload and underload more effectively. From the ROC curve it can be seen that the greater the area under ROC curve (the curve approaching the true Positive rate (TPR)) corresponds to lower threshold value. This means that all *PerClass* values being *underload* and *overload* exceeding that specific threshold are predicted as such. If such prediction made for a current class is correct, then the corresponding class value is true positive, otherwise it is false positive. Consequently, if for some values of the threshold the true positive rate (TPR) greatly exceeds the false positive rate (FPR), then the classification model with such threshold can be used to extract selectively the *underload* and *overload* class values from its mixture with the big number of *normal_load* values. Knowing the fact that the dataset is consisted of highly imbalanced class instances where *normal_load* values instances greatly supersedes the *overload* and *underload* values, the idea is to find the "predictive power" of every machine learning approach with respect to *overload* and *underload* class values. In order to find the optimal threshold for this "predictive power", we have applied a cost/benefit analysis. Cost benefit analysis gives the rate of the initial percentage of population (in our case data at hand) to predict the positive ratios of underload/overload of class values for the entire data. Figure 7 depicts the minimized cost benefit analysis for the machine learning approaches analyzed in this paper related to *overload* and *underload* class values.

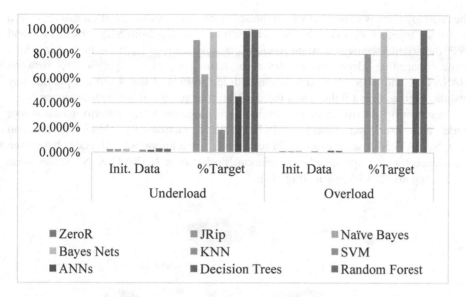

Fig. 8. The minimized cost/benefit analysis for the classifiers

The evaluation is done from the sense of percentage data which is required to predict a certain percentage of the overall target data. The results indicate that Rule Based learning models with JRip, Bayesian models with Naïve Bayes and Bayes Nets and Tree based models with Decision Trees and Random Forest show good prediction capability for more than 60% of the target data having less than 3% of initial data at hand. Interesting observations comes from tree-based models where we can see the production rules generated by the classifier that outlines the strong attributes in data used for classification. For this purpose we have observed two scenarios: the first scenario when

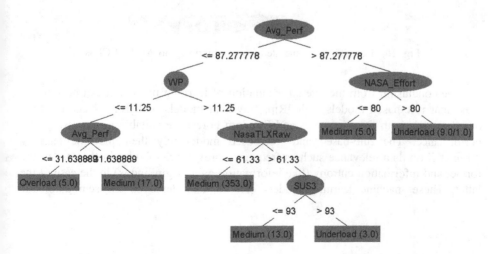

Fig. 9. Decision tree generated from scenario 1 (with AvgPerf Class)

the *AvgPerf* class is present in which the split criterion is determined on that class as a root node together with all other feature attributes coming from SUS, NASATLX and WP; the second scenario is when *AvgPerf* class is removed from the dataset in which some interesting dependencies between attributes arise. Specifically, between NASATLX and SUS which correlation outlined in Figs. 3 and 4 did not provide any insight. Figures 8 and 9 illustrates the decision tree generated for both scenarios.

In scenario 1 the overall classification task is based on Average Performance as root node and Workload Profile and NASATLX instruments (NASA_Effort and Raw NASATLSX measurements). In scenario 2, some important split nodes have risen from decision tree, specifically the relationship between NASATLX, SUS, Response Processing and NASA_Effort (Fig. 10).

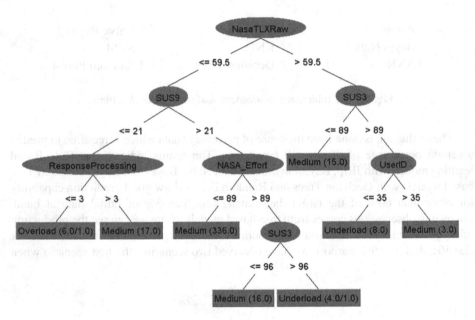

Fig. 10. Decision tree generated from scenario 2 (no AvgPerf Class)

The conclusion from the overall evaluation of learning models, it can be summarized that rule-based models with JRip, Bayesian models with BayesNets and tree-based models with Decision trees and Random Forest are suitable for assessing user performance. The rule-based and tree-based model rely their machine learning "strength" on data relevance such as frequency prevalence of rules in the case of the former and information entropy (like information gain or gini index) in the case of the latter. These machine learning models can be also deployed for semi-automatic detection of user performance by properly calculating the Workload profile (WP), System Usability Scale (SUS) and NASATLX instruments using the methodology elaborated in [22].

4   Conclusion and Future Work

The aim of this paper is to conduct a thorough evaluation of various supervised learning techniques for classifying performance underload/overload in web interfaces. The evaluation of classifiers is done by measuring the classifier accuracy, receiver operating characteristics (ROC) of each classifier and cost/benefit analysis. The results indicated that Rule-Based learning models with JRip, Bayesian models with Naïve Bayes and Bayes Nets and Tree based models with Decision Trees and Random Forest are the most robust ones.

Future work would involve:

- Enriching a dataset to create more evolutive models by using reinforcement learning approaches. This process of enrichment can be done by providing expert input on the dataset regarding classifier error while classifying performance overload/underload. It is obvious that none of the classification model training grasps all the aspects of user performance assessment. Incorporating the human factor in the process might improve the classification capability of the model when dataset become large, specifically when it involves semi-automatic user performance assessment.
- In this paper we have built classification models from subjective measures only without giving attention on psychological measures. These measures are looked with high interest by scientific community because of their intrinsic objectivity and the possibility of recording them without interfering with the main task. In this direction, considering hybrid machine learning models by combinin of subjective and physiological measures would be an approach worth exploring.
- The results when we used Artificial Neural Nets in our case presented poor classifier performance in assessing user performance with three given usability and MWL instruments. Exploring the Deep Learning as a machine learning technique would be an interesting research path. Since we have implied a wholistic approach on tackling user performance by using both usability (SUS) and MWL (NASATLX and WP) instruments and Deep Learning as a hierarchical learning method have a substantial *credit assignment path* (CAP) chain of transformations from input to output. This might potentially describe the causal relationship between input and output and, as a result, more deep relationships between the three instruments might emerge, thus increasing the classifier accuracy in total.

Appendix: Usability and Mental Workload Questionnaires and Independent Feature Descriptors

Table 6. System Usability Scale (SUS) questionnaire [22]

Label	Question
SUS_1	I think that I would like to use this interface frequently
SUS2	I found the interface unnecessarily complex
SUS_3	I thought the interface was easy to use
SUS_4	I think that I would need the support of a technical person to use this interface
SUS_5	I found the various functions in this interface were well integrated
SUS6	I thought there was too much inconsistency in this interface
SUS_7	I would imagine that most people would learn to use this interface quickly
SUS_8	I found the interface very unmanageable (irritating or tiresome) to use
SUS9	I felt very confident using the interface
SUS_{10}	I needed to learn a lot of things before I could get going with this interface

Table 7. NASA Task Load Index (NASATLX) questionnaire [22]

Label	Question
NT_1	How much mental and perceptual activity was required (e.g. thinking, deciding, calculating, remembering, looking, searching, etc.)? Was the task easy or demanding, simple or complex, exacting or forgiving?
NT_2	How much physical activity was required (e.g. pushing, pulling, turning, controlling, activating, etc.)? Was the task easy or demanding, slow or brisk, slack or strenuous, restful or laborious?
NT_3	How much time pressure did you feel due to the rate or pace at which the tasks or task elements occurred? Was the pace slow and leisurely or rapid and frantic?
NT_4	How hard did you have to work (mentally and physically) to accomplish your level of performance?
NT_5	How successful do you think you were in accomplishing the goals, of the task set by the experimenter (or yourself)? How satisfied were you with your performance in accomplishing these goals?
NT_6	How insecure, discouraged, irritated, stressed and annoyed versus secure, gratified, content, relaxed and complacent did you feel during the task?

Table 8. The Workload Profile (WP) questionnaire [22]

Label	Question
WP_1	How much attention was required for activities like remembering, problem-solving, decision-making, perceiving (detecting, recognising, identifying objects)?
WP_2	How much attention was required for selecting the proper response channel (manual - keyboard/mouse, or speech - voice) and its execution?
WP_3	How much attention was required for spatial processing (spatially pay attention around you)?
WP_4	How much attention was required for verbal material (e.g. reading, processing linguistic material, listening to verbal conversations)?
WP_5	How much attention was required for executing the task based on the information visually received (eyes)?
WP_6	How much attention was required for executing the task based on the information auditorily received (ears)?
WP_7	How much attention was required for manually respond to the task (e.g. keyboard/mouse)?
WP_8	How much attention was required for producing the speech response (e.g. engaging in a conversation, talking, answering questions)?

Table 9. Data descriptors for the five independent feature sets from the data

Independent feature	N	Mn. (μ)	Std (σ)	Var. (s)	Kurtosis	Skewness	Min	Max
Feature Set #1 – NASATLX Questions								
NASA_Mental	405	50.76	26.82	719.11	−1.05	0.24	−0.25	0.12
NASA_Temporal	405	39.54	29.8	887.78	−1.12	0.24	0.34	0.12
NASA_Effort	405	56.38	25.75	663.21	−0.72	0.24	−0.53	0.12
NASA_Performance	405	67.95	29.41	864.8	−0.12	0.24	−0.94	0.12
NASA_PsychologicalStress	405	37.17	29	841.27	−0.9	0.24	0.52	0.12
NASA_Mental	405	50.76	26.82	719.11	−1.05	0.24	−0.25	0.12
Feature Set #3 – Pairwise comparison of NASATLX								
NASA_MenTem	405	0.37	0.48	0.23	−1.69	0.56	0	1
NASA_MenPsy	405	0.3	0.46	0.21	−1.2	0.9	0	1
NASA_MenEff	405	0.63	0.48	0.23	−1.73	−0.53	0	1
NASA_MenPer	405	0.51	0.5	0.25	−2.01	−0.02	0	1
NASA_TemPsy	405	0.42	0.49	0.24	−1.9	0.34	0	1
NASA_TemEff	405	0.63	0.48	0.23	−1.73	−0.53	0	1
NASA_TemPer	405	0.62	0.48	0.24	−1.74	−0.52	0	1
NASA_PsyEff	405	0.73	0.45	0.2	−0.94	−1.03	0	1
NASA_PsyPer	405	0.71	0.46	0.21	−1.18	−0.91	0	1
NASA_EffPer	405	0.52	0.5	0.25	−2	−0.07	0	1

(continued)

Table 9. (*continued*)

Independent feature	N	Mn. (μ)	Std (σ)	Var. (s)	Kurtosis	Skewness	Min	Max
Feature Set #3 Total Preferences of pairwise Comparison (weighted NASATLX)								
NASA_menTotPref	405	3.2	1.13	1.27	−0.71	−0.23	1	5
NASA_TemTotPref	405	2.7	1.37	1.88	−1.11	0.36	1	5
NASA_PsychTotPref	405	2.28	1.34	1.8	−0.72	0.71	1	5
NASA_EffTotPref	405	3.46	1.1	1.21	−0.63	−0.36	1	5
NASA_PerTotPref	405	3.36	1.31	1.71	−1.05	−0.27	1	5
Feature Set #4 The original Workload Profile (WP)								
CentralProcessing	405	53.02	27.36	748.37	−0.95	−0.36	0	100
ResponseProcessing	405	33.92	27.14	736.74	−0.96	0.49	0	100
SpatialProcessing	405	23.97	24.34	592.63	0.27	1.06	0	100
VerbalProcessing	405	51.59	34.43	1185.47	−1.42	−0.22	0	100
VisualInput	405	62.24	27.58	760.42	−0.48	−0.66	0	100
AuditoryInput	405	33.25	37.78	1427.43	−1.23	0.67	0	100
ManualResponse	405	30.18	26	675.79	−0.68	0.62	0	100
SpeechResponse	405	12.06	18.28	334.17	3.85	1.98	0	100
Feature Set #5 The original System usability Scale (SUS)								
SUS1	405	64.73	29.15	849.77	−0.36	−0.8	0	100
SUS2	405	26.73	25.81	666.22	0.68	1.2	0	100
SUS3	405	70.6	26.2	686.36	0.39	−1.08	0	100
SUS4	405	11.63	15.99	255.83	7.56	2.42	0	100
SUS5	405	65.77	25	624.99	−0.04	−0.74	0	100
SUS6	405	24.78	24.58	604.06	0.94	1.23	0	100
SUS7	405	72.13	25.57	653.83	0.68	−1.2	0	100
SUS8	405	26	28.01	784.29	0.48	1.24	0	100
SUS9	405	72.28	24.98	623.95	0.7	−1.14	0	100
SUS10	405	14.07	17.64	311.22	4.3	1.94	0	100

References

1. Reid, G.B., Nygren, T.E.: The subjective workload assessment technique: a scaling procedure for measuring mental workload. J. Adv. Psychol. **52**, 158–218 (1988)
2. Stassen, H.G., Johannsen, G., Moray, N.: Internal representation, internal model, human performance model and mental workload. J. Autom. **26**(4), 811–820 (1990)
3. Wickens, C.D.: Mental workload: assessment, prediction and consequences. In: Longo, L., Leva, M.C. (eds.) H-WORKLOAD 2017. CCIS, vol. 726, pp. 18–29. Springer, Cham (2017). https://doi.org/10.1007/978-3-319-61061-0_2
4. Longo, L.: A defeasible reasoning framework for human mental workload representation and assessment. Behav. Inf. Technol. **34**(8), 758–786 (2015)

5. Longo, L.: Human-computer interaction and human mental workload: assessing cognitive engagement in the world wide web. In: Campos, P., Graham, N., Jorge, J., Nunes, N., Palanque, P., Winckler, M. (eds.) INTERACT 2011. LNCS, vol. 6949, pp. 402–405. Springer, Heidelberg (2011). https://doi.org/10.1007/978-3-642-23768-3_43

6. Longo, L.: Formalising human mental workload as non-monotonic concept for adaptive and personalised web-design. In: Masthoff, J., Mobasher, B., Desmarais, Michel C., Nkambou, R. (eds.) UMAP 2012. LNCS, vol. 7379, pp. 369–373. Springer, Heidelberg (2012). https://doi.org/10.1007/978-3-642-31454-4_38

7. Longo, L.: Designing medical interactive systems via assessment of human mental workload. In: International Symposium on Computer-Based Medical Systems, pp. 364–365 (2015)

8. Moustafa, K., Luz, S., Longo, L.: Assessment of mental workload: a comparison of machine learning methods and subjective assessment techniques. In: Longo, L., Leva, M.Chiara (eds.) H-WORKLOAD 2017. CCIS, vol. 726, pp. 30–50. Springer, Cham (2017). https://doi.org/10.1007/978-3-319-61061-0_3

9. Blankertz, B., Curio, G., Muller, K.R.: Classifying single trial EEG: towards brain computer interfacing. In: Advances in Neural Information Processing Systems, vol. 1, pp. 157–164 (2002)

10. Dornhege, G., Blankertz, B., Curio, G., Muller, K.R.: Boosting bit rates in noninvasive EEG single-trial classifications by feature combination and multiclass paradigms. IEEE Trans. Biomed. Eng. **51**(6), 993–1002 (2004)

11. Stevens, R., Galloway, T., Berka, C.: Integrating EEG models of cognitive load with machine learning models of scientific problem solving. In: Proceedings of 2nd Annual Augmented Cognition International Conference, pp. 55–65 (2006)

12. Zhang, Y.Z.Y., Owechko, Y., Zhang, J.Z.J.: Driver cognitive workload estimation: a data-driven perspective. In: Proceedings of the 7th International IEEE Conference on Intelligent Transportation Systems (IEEE Cat. No.04TH8749), pp. 642–647 (2004)

13. Lee, J.C., Tan, D.S.: Using a low-cost electroencephalograph for task classification in HCI research. In: Proceedings of the 19th ACM Symposium on User Interface Software and Technology, pp. 81–90 (2006)

14. Reid, G.B., Nygren, T.E.: The subjective workload assessment technique: a scaling procedure for measuring mental workload, vol. 52, pp. 185–218. North-Holland (1988)

15. Hart, S.G., Staveland, L.E.: Development of NASA-TLX (task load index): results of empirical and theoretical research. In: Advances in Psychology, vol. 52, pp. 139–183 (1988)

16. Kramer, A.F.: Physiological metrics of mental workload: a review of recent progress. Multiple-task performance. Taylor & Francis, 279–328 (1991)

17. Wickens, C.D.: Multiple resources and mental workload. Hum. Factors **50**, 449–454 (2008)

18. Wickens, C.D., Hollands, J.G.: Engineering Psychology and Human Performance, 3rd edn. Prentice Hall, Upper Saddle River (1999)

19. Tsang, P.S., Velazquez, V.L.: Diagnosticity and multidimensional subjective workload ratings. Ergonomics **39**(3), 358–381 (1996)

20. Brooke, J.: SUS-A quick and dirty usability scale. Usability Eval. Ind. **189**(194), 4–7 (1996)

21. Azevedo, A.I R.L., Santos, M.F.: KDD, SEMMA and CRISP-DM: a parallel overview. IADS-DM (2008)

22. Longo, L., Dondio, P.: On the relationship between perception of usability and subjective mental workload of web interfaces. In: 2015 IEEE/WIC/ACM International Conference on Web Intelligence and Intelligent Agent Technology (WI-IAT), pp. 345–352. IEEE (2015)

23. Liang, N.Y., Saratchandran, P., Huang, G.B., Sundararajan, N.: Classification of mental tasks from EEG signals using extreme learning machine. Int. J. Neural Syst. **16**(01), 29–38 (2006)

24. Müller, K.R., Tangermann, M., Dornhege, G., Krauledat, M., Curio, G., Blankertz, B.: Machine learning for real-time single-trial EEG-analysis: from brain–computer interfacing to mental state monitoring. J. Neurosci. Methods **167**(1), 82–90 (2008)

25. Yin, Z., Zhang, J.: Identification of temporal variations in mental workload using locally-linear-embedding-based EEG feature reduction and support-vector-machine-based clustering and classification techniques. Comput. Methods Programs Biomed. **115**, 119–134 (2014)

26. Zhang, J., Yin, Z., Wang, R.: Recognition of mental workload levels under complex human–machine collaboration by using physiological features and adaptive support vector machines. IEEE Trans. Hum.-Mach. Syst. **45**(2), 200–214 (2014)

27. Yin, Z., Zhang, J., Wang, R.: Neurophysiological feature-based detection of mental workload by ensemble support vector machines. In: Wang, R., Pan, X. (eds.) Advances in Cognitive Neurodynamics (V). ACN, pp. 469–475. Springer, Singapore (2016). https://doi.org/10.1007/978-981-10-0207-6_64

28. Rubio, S., Díaz, E., Martín, J., Puente, J.: Evaluation of subjective mental workload: a comparison of SWAT, NASA-TLX, and workload profile methods. Appl. Psychol. **53**, 61–86 (2004)

29. Rani, P., Liu, C., Sarkar, N., Vanman, E.: An empirical study of machine learning techniques for affect recognition in human–robot interaction. Pattern Anal. Appl. **9**, 58–69 (2006)

30. Smith, K.T.: Observations and issues in the application of cognitive workload modelling for decision making in complex time-critical environments. In: Longo, L., Leva, M.Chiara (eds.) H-WORKLOAD 2017. CCIS, vol. 726, pp. 77–89. Springer, Cham (2017). https://doi.org/10.1007/978-3-319-61061-0_5

31. Balfe, N., Crowley, K., Smith, B., Longo, L.: Estimation of train driver workload: extracting taskload measures from on-train-data-recorders. In: Longo, L., Leva, M. (eds.) H-WORKLOAD 2017. CCIS, vol. 726, pp. 106–119. Springer, Cham (2017). https://doi.org/10.1007/978-3-319-61061-0_7

32. Cahill, J., et al.: Adaptive automation and the third pilot: managing teamwork and workload in an airline cockpit. In: Longo, L., Leva, M. (eds.) H-WORKLOAD 2017. CCIS, vol. 726, pp. 161–173. Springer, Cham (2017). https://doi.org/10.1007/978-3-319-61061-0_10

33. Delamare, L., Golightly, D., Goswell, G., Treble, P.: Quantification of rail signaller demand through simulation. In: Longo, L., Leva, M.Chiara (eds.) H-WORKLOAD 2017. CCIS, vol. 726, pp. 174–186. Springer, Cham (2017). https://doi.org/10.1007/978-3-319-61061-0_11

34. Byrne, A.: Mental workload as an outcome in medical education. In: Longo, L., Leva, M. Chiara (eds.) H-WORKLOAD 2017. CCIS, vol. 726, pp. 187–197. Springer, Cham (2017). https://doi.org/10.1007/978-3-319-61061-0_12

Operator Functional State: Measure It with Attention Intensity and Selectivity, Explain It with Cognitive Control

Alexandre Kostenko[1]([⊠]), Philippe Rauffet[2], Sorin Moga[1], and Gilles Coppin[1]

[1] Lab-STICC, UMR CNRS 6285, IMT Atlantique, Brest, France
{alexandre.kostenko,sorin.moga,
gilles.coppin}@imt-atlantique.fr
[2] Lab-STICC, UMR CNRS 6285, Université Bretagne Sud, Lorient, France
philippe.rauffet@univ-ubs.fr

Abstract. To improve the safety and the performance of operators involved in risky and demanding missions, human-machine cooperation should be dynamically adapted, in terms of dialogue or function allocation. To support this reconfigurable cooperation, a crucial point is to assess online the operator's ability to keep performing the mission, to anticipate and predict potential future performance impairments, as well as to be able to activate appropriate countermeasures in time. Thus, the paper explores the concept of Operator Functional State (OFS) developed by Hockey in 2003, by articulating it with underlying cognitive and attentional states, as well as with the notion of cognitive control modes.

Keywords: Operator Functional State · Mental effort · Visual attention · Cognitive control modes

1 Introduction

Many operators carry out their activity in complex, high-risk situations and with strong time pressure. This is the case in air domain, for fighter pilots [1, 2] or for drone operators [3–6]. In this context, an important challenge is the improvement of the safety and the performance of the operators. This challenge could be solved by adjusting in real time the dialogue and the cooperation between man and machine according to the state of the human operator [7, 8]. It therefore becomes crucial to assess online the operator's ability to keep performing the mission, to anticipate and predict potential future performance impairments, as well as to be able to activate appropriate countermeasures in time (change in informational transparency level, dynamic allocation of functions, etc.). Many research works have already addressed this question of modelling and assessing operator functional state [9–11], either by focusing on mental workload, stress, fatigue, engagement, etc. These different cognitive states, usually measurable with behavioural and neurophysiological indicators, are promising for determining online if an operator is capable - or not - to perform tasks. Nevertheless,

L. Longo and M. C. Leva (Eds.): H-WORKLOAD 2019, CCIS 1107, pp. 156–169, 2019.
https://doi.org/10.1007/978-3-030-32423-0_10

they generally produce only binary information (stressed or not, fatigued or not, etc.). Moreover, if they are often considered in the literature as risk indicators for performance degradation, they can however produce "false positives". For instance, a high level of mental workload might not always be deleterious for the operator and that could sometimes reveal that he/she is facing a new, complex problem; in this case, it could be counterproductive to react to this high mental workload, and to interfere with operator's activity by reconfiguring the system or triggering distracting alarms. To overcome these limits of binarization and false positivity risk in the evaluation of operator functional state (OFS), an idea would be to use the notion of cognitive control strategies or modes, developed by different authors [6, 9, 12] as compensatory or regulatory mechanisms to dynamically adjust performance and cognitive cost. This framework could help for better characterizing the OFS, and would improve the relevance and the accuracy of system adaptation and assistance provided to operators facing dynamic and complex situations. This paper aims at exploring the articulation between the notion of operator functional state, the attention theory and the framework of cognitive control. First, a literature review provides the definitions of these elements and emphasizes the conceptual relationships between them. A methodological framework is then developed, to present the useful neurophysiological and behavioural indicators as well as an architecture to classify these indicators into finer-grained categories of operator functional states, labelled with cognitive control modes. Finally, this approach is discussed in terms of theoretical and operational perspectives.

2   Theoretical Background: Articulating OFS Concept with Attention and Cognitive Control Modes

This part defines the notion Operation Functional States (OFS), then presents two approaches to operationally estimate this concept. On the one hand, there is a data-driven approach aiming at modelling and assessing OFS as a risk of performance decrement, based on the neurophysiological and behavioral measurement of cognitive states like mental effort, mental fatigue, or selective attention. On the other hand, a model-driven approach could use the conceptual framework of cognitive control modes to distinguish and label different classes of the OFS.

2.1   The Concept of Operator Functional State

To encapsulate the different elements contributing to a potential degradation of performance from an operator, Hockey [10] proposes the notion of Operator Functional State. This concept is defined as *"the variable capacity of the operator for effective task performance in response to task and environmental demands, and under the constraints imposed by cognitive and physiological processes that control and energise behaviour."* This definition first underlines a strong relationship between the operator functional state and his/her performance on the tasks, leading to OFS classification using categories like "Capable/Incapable", "Low risk/Moderately risky/Very risky", or more generally classes expressing a gap to expected performance. However, it is often difficult to predict a performance collapse of an operator solely on the basis of the analysis of the results of his/her activity. This difficulty is particularly pregnant for experienced operators: the observable degradations of their performance are indeed only slight and gradual (before a stall), because these operators have regulatory strategies to maintain during a certain time the effectiveness of the main tasks. Therefore, the time of reaction and adaptation of the system to a performance collapse may be too long to be caught for the situation and may thus cause irreversible effects [10]. Thus, the OFS concept aims at coping with these difficulties for anticipating a decrease in the performative capacity of operator. According to Hockey [10], OFS can be attached to underlying cognitive states. The overall OFS is especially associated to mental workload and stress [10, 13], but is also related to engagement, alertness and fatigue, or situation awareness [11]. These different cognitive states are useful to operationalize this abstract OFS concept. Indeed, most of these states are quantifiable, based on physiological and behavioural indicators referring to the cognitive activities produced by operators [1, 14], relatively to some context indicators giving the situation requirements [15, 16]. These cognitive states, a kind of intermediate level of the overall OFS, will make possible to distinguish a sufficient capacity (functional state) from an insufficient capacity (non-functional state) to carry out the task.

2.2   Modelling and Assessing OFS with Attention Intensity and Selectivity

The underlying cognitive states mainly investigated in the literature about OFS can be defined according the attention theory, and especially the model proposed by Van Zomeren and Brouwer [18]. Attention is schematized into two main components: a resource-based view of attention, named intensity, and a process-based view of attention, called selectivity. Alertness and sustained attention (or vigilance) were incorporated into intensity, while selectivity considered mechanisms and processes about focused attention (visual search, overt attention, distractor inhibition) and divided attention (in multitasking). Adopting this model, three main cognitive states are generally considered to assess the OFS (Fig. 1).

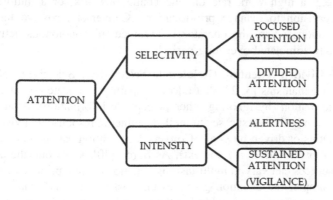

Fig. 1. Attention components [17]

Mental Effort (Related to Alertness). Alertness can be decomposed into two categories, namely tonic and phasic components. Tonic alertness refers to internal control and the maintenance of arousal to respond to events in the environment. As a primitive component of attention, it reflects the activation of the organism at any time [18]. Phasic alertness refers to a short-lived effect of responding to a salient external event [19]. As emphasised by Valdez et al. [21], tonic alertness varies according to circadian variations, and is impaired by sleep deprivation. Conversely, phasic alertness is influenced by residual task-switching cost [21], and high attention-demanding tasks [22]. The variations in tonic and phasic alertness can then result in the production of a mental effort. Mulder [24] discriminates between two types of mental effort, i.e. the mental effort devoted to the processing of information in controlled mode (computational effort) and the mental effort needed to apply when the operator's energetical state is affected (compensatory effort). Computational effort is exerted to maintain phasic alertness, by keeping task performance at an acceptable level, for instance, when task complexity level varies or secondary tasks are added to the primary task. Compensatory effort takes care of performance decrement in case of tonic alertness reduction up to a certain level.

Mental Fatigue (Related to Sustained Attention). Sustained cognitive activity can cause a state of mental fatigue characterized by a decrease in alertness and mental effort over time, especially in the case of an operator performing complex mental tasks or dealing with stressful events [24]. Desmond and Hancock [26] and Gimeno et al. [27] make a distinction between active fatigue - which concerns the reduction in the commitment of the task after an intense work, and can be characterized by the choice of strategies with less effort [27] - and passive fatigue - which corresponds to a loss of alertness caused by low operator stimulation. In addition to the time on task factor (which allows us to characterize the sustainability of mental effort), active fatigue can be exacerbated by poor conditions of information visibility or accessibility ("blind" piloting in the case of aircraft navigating in night or bad weather conditions, where the environment is not directly observable), but also by the complexity and the continuity of the task (e.g. a high work rate on one continuous task, or a multitask activity requiring supervision to manage prioritization). Conversely, passive fatigue corresponds to situations where the operator is involved in monotonous activity, where he/she is weakly stimulated over long periods.

Focused and Divided Attention (Related to Selectivity). Selectivity refers to cognitive control mechanisms [3, 28] that allow to sparingly allocate cognitive resources on certain information (by ignoring other perceptible information). Attention can be "focused", or even in the worst case "tunnelled", either in response to a visual, auditory or spatial stimulus or driven by goals. Conversely, attention can also be "divided" (to keep control over a multitasking context). As Bardy [30] points out, the allocation of attentional resources can vary in multi-tasking situations with a priority task. When the main task is simple, the attention given to this task is low and a large amount of attentional resources remain available for the secondary task (allowing divided attention). On the contrary, when the main task is difficult, the resources available for the secondary task are reduced, as well as the performance of the operator (corresponding to a focused attention). Finally, when the difficulty is very strong, the secondary task is set aside (with a phenomenon of tunneling).

Modelling and estimating the OFS from these kinds of cognitive states seems interesting, since these ones are measurable within a data-driven approach, either with neurophysiological indicators for assessing attention intensity, either by behavioural indicators to address attention selectivity. These indicators will be further explored in Sect. 3.1. Nevertheless, it could be difficult to determine precisely from only these cognitive states if an operator is really functional or not (with the risk of false positivity and the limit of OFS binarization mentioned above). This data-driven approach could therefore be combined with a model-driven framework, to identify OFS finer-grained categories encapsulating regulatory or compensatory mechanisms implemented by operators.

2.3   Classifying and Explaining OFS with the Notion of Cognitive Control

In his work on the OFS, Hockey [10] also defines a compensatory control model, which explains how the operator can implement a self-regulatory mechanism to stay functional.

Compensatory Control Modes. The self-regulatory mechanism defined by Hockey [10] corresponds to the selection of an adaptive response mode to cope with a strong task demand, with consequences on both the performance and the cognitive states defined above. These modes are as follows:

- *Engaged:* the operator applies a high level of control over the task, with an optimal performance. This mode produces increased mental effort, and decreased fatigue.
- *Strain:* the operator gets an adequate performance, but with a low level of control. The effort produced by operator comes with by an increase in anxiety and fatigue.
- *Disengaged:* This mode corresponds to a deleterious regulation, with a decrease in performance. The operator produces a lower mental effort, but this is accompanied by an increase in anxiety and the maintenance of the pre-existing level of fatigue.

This notion of control loops or modes is also considered by other authors, that proposed models also relating operator's cognitive control with the level of performance and the effects on different cognitive states (mental effort, fatigue, selective attention), in situations of low or high task demand.

Engagement Critical Modes. Dehais et al. [31] proposed therefore to distinguish 3 critical states of engagement completing the modes of Hockey: *dis-engagement*, active or passive, which can here correspond either to a mental exhaustion or to an attentive wander; *over-engagement*, which can produce attentional tunneling; *in-engagement*, which leads to erratic behavior, where the operator can no longer focus on critical information.

Cognitive Control Modes (COCOM). Moreover, Hollnagel [13] also proposed a typology of control modes, distinguishing 5 levels, ranging from the most reactive (*scrambled, opportunistic and tactical unattended modes*) to the most proactive ones (*tactical attended and strategic modes*). It is established that the performance is low for the most reactive modes while it is satisfactory for the most proactive modes. The instantiation of the modes by the operators depends on the familiarity of the situation (and therefore the expertise of the operators), as well as external constraints (in particular the number of data, the number of simultaneous tasks, and - especially - the available time).

Mental Workload Regulation Loops. Finally, Kostenko et al. [7], modelled regulation loops of mental workload, considered with a capacity view as a balance between task demand and operator's cognitive capacity (cf. [31]). These authors particularly pointed out the existence of two mechanisms of task simplification, lowering the cognitive cost or the task constraints with maintaining an acceptable but degraded performance: a *priority regulation* aimed at prioritizing certain main tasks to the detriment of secondary tasks; a *efficiency regulation* is implemented by operators to keep attention shared on the whole of the tasks, but with a slight decrement in performance or processing time.

A synthesis of the definition and the sources of these cognitive control modes (CCM) is presented in Table 1. Table 2 summarizes the relationships between these modes and the resulting performance, the contextual task demand or constraint (relative to operator's capacity), and the three different cognitive states investigated in this paper (mental effort, fatigue and selective attention).

Table 1. Cognitive control modes (CCM) as labels for OFS classes

OFS categories labelled with cognitive control modes		Description	Authors
Strategic		Operator has time anticipate situation and elaborate or plan an optimal response	[12]
Tactical	Tactical (attended)	Operator carefully follows a procedure with a high control	[12]
	Tactical (unattended)	Operator carries out a routine, with a low control	[6, 12]
Strain		Operator produce a supplementary and costly effort to maintain a satisfying level of performance	[6, 9]
Simplification	Priority	Operator gives priority on some main tasks to the detriment of secondary tasks	[6, 9, 30]
	Efficiency	All the tasks are performed, but with reduced effectiveness or longer processing time	[6, 9]
Opportunistic		Operator reacts to stimuli, with a very partial view of the situation	[12]
Scrambled		Operator tries many actions erratically, without understanding the situation	[12]
In-engagement		Operator is unable to focus on a task or to react to a critical information	[12, 30]
Disengagement	Active disengagement	Operator is mentally exhausted	[9, 30]
	Passive disengagement	Operator is bored	[30]

Table 2. Relationships between CCM

OFS categories labelled with cognitive control modes	Performance	Demand	Intensity: mental effort	Intensity: fatigue	Selectivity
Strategic	Optimal	<Capacity	Medium-High	Low	Divided
Tactical (unattended)	Optimal	<Capacity	Low	Low	Divided
Tactical (attended)	Optimal	=Capacity	Medium-High	Low	Divided
Strain	Adequate	>Capacity	High	Medium	Divided
Priority	Adequate	>Capacity	Medium-High	Medium	Focused on goals (tunneling risk)
Efficiency	Adequate	>Capacity	Medium	Medium	Divided
Opportunistic	Impaired	>Capacity	High	High	Focused on stimuli (tunneling risk)
Scrambled	Impaired	>Capacity	High	High	Divided
In-engagement	Impaired	>Capacity	High	High	Divided but without focus
Active disengagement	Impaired	>Capacity	Low	High	–
Passive disengagement	Impaired	<Capacity	Low	High (Passive)	–

These different CCM could be used to label different classes of the OFS. That would allow to distinguish between:

- *deleterious OFS classes:* these ones result in impaired performance and it is necessary to provide a prompt assistance to the operator;
- *risky OFS classes:* these OFS categories result in adequate performance but are characterized by a task demand higher than operator capacity. It is therefore important to carefully monitor the evolution of the OFS. Sometimes, if the intervention does not interfere with the efficient compensatory strategies carried by operators, some very fine-tuned assistance can be also provided to the operator (e.g. with visual cues, or partial delegation of some sub-tasks to the machine).

3   Methodological Framework: Proposal of Architectures to Classify Cognitive States and CCM-Related OFS

This part presents the input data and classification architectures that could support the detection of the cognitive states (mental effort, fatigue, selective attention) and the different levels of OFS labelled in terms of CCM.

3.1   Neurophysiological and Behavioral Indicators as Inputs for OFS Classification

The different cognitive states, composing the OFS, are usually evaluated by the measurement and the observations of several type of variables: we can distinguish neuro-physiological, behaviour and context indicators.

Neuro-Physiological Indicators (to Approach Attention Intensity and Assess Mental Effort and Fatigue). Since Kahneman's energetic approach [32] emphasizing the relationship between physiological activation and mental activity, many cognitive states have been associated with variations in physiological signals (Heart Rate HR or Heart Rate Variability HRV, electrodermal activity, pupillary diameter, etc.). Moreover, with Parasuraman [34] and the rise of neuro-ergonomics, new approaches are focusing on the link between states of the operator and the waves generated by brain activity (with devices such as EEG or fNIRS, cf. [34]). Physiological signals allow to detect the cognitive states. They are generally not used in raw form, but they initially produce indicators that are then processed by classification algorithms. These indicators are either directly derived from the measurement (as for example the heart rate established by simple observation of the QRS complex) or built after a projection of the measurements in a dedicated representation space (wavelets, etc. …). It may also be necessary to incorporate a data cleansing step (e.g. removing RED-NS parasites for electrodermal activity) and normalization of the data against a baseline [2]. The signals relating to central nervous system (measured with Electroencephalography or Functional Near-Infrared Spectroscopy, namely EEG or FNIRS) are very sensitive and very discriminating on the different operating states [35], but the sensors are for the moment very invasive, and the treatments are relatively complex (aggregation by fusion of data of the different brain regions in temporal and frequency domains), which can limit the real-time application of these techniques. Nevertheless, the recent development of fNIRS seems a promising perspective: this technique, less invasive and limited to the prefrontal cortex, proposes a smaller number of variables that can be associated with cognitive or emotional states (oxyhemoglobin HB02, deoxyhemoglobin HBB, cerebral blood volume CBV or cerebral oxygen exchange COE [34, 36]). The physiological signals dependent on autonomic nervous system (heart rate, electrodermal activity, pupillary diameter) make it possible to approach the sympathetic (activation) and parasympathetic (awakening at rest) tendencies of the organism. They are therefore sensitive to mental effort and fatigue. On the other hand, they have a weak diagnosticity: it is indeed difficult to discriminate using only physiological signals from cognitive states of the same tendency (for example stress, attentional focus or mental workload). It should also be noted that the electrodermal activity is not sensitive enough for short-term treatment: the signal tends to react quickly to a stimulus but decreases slowly when the stimulus disappears [14]. The latency effect is therefore too important to monitor the OFS in real time.

Behavioural Indicators (to Approach Attention Selectivity). Some authors also propose integrating strategies into the assessment of cognitive states, using markers derived from the observation of operator behavior [6, 37]. The ocular indicators are especially an interesting marker of operator behavior, related to visuospatial research activities. Indeed, the signals related to somatic nervous system, and particularly the eye movements, make it possible to confirm certain states, such as fatigue with blinks or PERCLOS (namely percentage of eye closure), or attentional focus with saccadic rate and gaze entropy [38]. When contextualized with the scene or the area of interests, they also enable to better understand certain phenomena (attentional focus, perseverance, etc.) indicating the operators' visual workload and strategies [39]. There is therefore no ideal solution with a single measurement able to indicate the operator state. Each solution faces a problem of invasiveness, sensitivity, selectivity or diagnosticity [31, 40]. Real-time monitoring and characterization of the OFS should therefore rely on the combination and fusion of physiological indicators (DP, HR signals) and behavioral indicators (ocular activity).

Context Indicators. In addition, the weak diagnosticity of some of these data tends to show the need to contextualize the physiological indicators to better determine and categorize the OFS. To face this challenge, several research works propose to "situate" the cognitive states according to the constraints of the situation and the level of task requirement [6, 16, 41].

As a summary, we propose the following summary table by pointing out the relevant neuro-physiological and behavioural indicators, and by distinguishing them according to their sensitivity to the different cognitive states.

Table 3. Indicators related to cognitive states (↑: increase, ↓: decrease, -: unsensitive), adapted from Lassalle et al. (2016)

Categories of indicators	Indicators	Intensity: mental effort	Intensity: fatigue	Selectivity
Neuro-physiological markers	HBO2, CBV, COE	↑	-	-
	HBB	↓	-	-
	HR	↑	-	-
	HRV	↓	-	-
	Pupil diameter	↑	-	-
Behavioral markers (ocular)	Fixation duration	↑	-	↓ when important visual search and divided attention ↑ when focusing or tunneling
	Saccadic rate	↓	↓	↑ when important visual search and divided attention ↓ when focusing or tunneling
	Gaze entropy, Fixation dispersion	-	-	↑ when important visual search and divided attention ↓ when focusing or tunneling
	Blink rate	-	↑	-
	PERCLOS	-	↑	-

3.2 Classification Architecture

The objective of the classification is twofold: detecting from neurophysiological signals and visual behaviour different cognitive states (mental effort, fatigue and selective attention, cf. Fig. 2 blue boxes), but also estimating the Operator Functional State according to different Cognitive Control Modes (effective or not, reactive or proactive, mentally demanding or not, cf. Fig. 2 yellow box). To carry out these multiple classifications (cf. Fig. 2 flags), the kinds of indicators synthesized in Table 3 are generally trained with contextual labels (relative to the level of difficulty of the task to be performed), subjective assessment (NASA-TLX or the Instantaneous self-assessment scores for mental effort), or with performance evaluation (like the Psycho motor Vigilance Task for fatigue or double task performance for mental effort). Moreover, the cognitive control modes are usually labelled according to three types of performance indicators, indicating the adequate or deleterious effect, as well as the proactive or reactive nature of the modes. There are markers related to the task effectiveness (number of errors, or gaps relatively to the objectives), markers related to the time of reaction (or remaining time budget), and markers related to the actions carried out by operators to perform the task (e.g. action rate on the actuators).

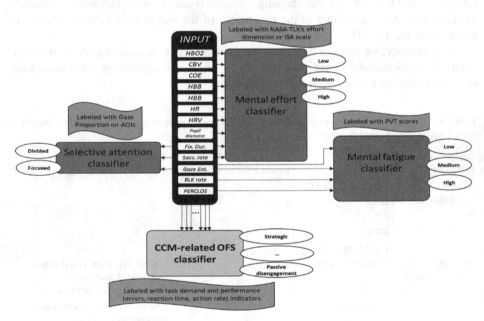

Fig. 2. Classification architecture for cognitive states and OFS modes (Color figure online)

Recent works related to the classification of mental effort and fatigue [11, 42, 43] have shown that supervised learning can be effectively used to detect different levels of cognitive states from physiological indicators, using different classifiers (Support Vector Machines, k-Nearest Neighbors, Random Forest algorithms). These first results, however, were based on a very discrete classification (2 or 3 maximum classes for each cognitive state), and could be further investigated, with the notion of cognitive control modes as classes for OFS.

4   Concluding Discussion

This paper proposes to model and operationalize the combination of two approaches, a data-driven approach based on neurophysiological and ocular indicators and a model-driven approach based on cognitive control modes, to refine the detection and classification of the OFS. This proposal brought both theoretical and practical contributions and perspectives.

Theoretical Contributions and Perspectives. On the one hand, the data-driven approach proposes to associate the state of use of cognitive resources (mental effort, fatigue) with indicators on visual attentional behavior (selectivity of attention). This association would avoid a false positivity in detecting a deleterious level of OFS. Indeed, a high mental effort or fatigue can induce a shortage or a lack of resources, but this can also result from the implementation of a virtuous compensatory mechanism from the operator. Integrating the visual search behavior would therefore lead to a robust evaluation of the OFS; Thus, if a strong mental effort is also associated with a less performing visual search in a multitasking context, it will be possible to confirm the detection of a deleterious level of OFS. On the other hand, the use of cognitive control modes would make it possible to better distinguish the OFS levels and to exceed the binarization limits of the classification. We could thus discriminate between more or less costly performance modes, and deleterious modes with a variable cognitive cost.

Practical Contributions and Perspectives. This theoretical framework, as well as the classification architecture proposed in this paper, will be implemented in an experimental setting, on the MATB-II simulator. Data will thus be collected from participants carrying out a scenario. This scenario will be designed by varying the multi-task context (to bring different types of selective attention), the type of main task (leading a more or less proactive behavior such as tank management, or reactive like tracking management)., and the level of difficulty within each experimental condition (causing a mental effort and more or less significant fatigue). We then aim to classify the behavioral and neurophysiological data collected, in terms of cognitive states (mental effort, fatigue and selective attention), but also in terms of classes of the OFS, labeled in terms of CCM. The obtained granularity of the OFS would also allow to contextualize and specify the assistance to be given to an operator in a complex or even critical situation. We could thus distinguish on the one hand the emergency responses to be implemented against the detection of critical control modes leading to degraded performance, and on the other hand the proportionate responses to be provided for risky modes with adequate performance but costly cognitively. In this last case, the aided provided to the operator should avoid triggering unnecessary countermeasures or even interfering with the mechanisms of regulation or compensation operators.

Acknowledgement. The authors would like to thank the DGA (Direction Générale de l'Armement), Thales AVS and Dassault Aviation which support the funding of this study and the scientific program "Man-Machine Teaming" in which the research project PRECOGS occurs.

References

1. Veltman, J.A., Gaillard, A.W.K.: Physiological indices of workload in a simulated flight task. Biol. Psychol. **42**(3), 323–342 (1996)
2. Lassalle, J., et al.: COmmunication and WORKload analyses to study the COllective WORK of fighter pilots: the COWORK2 method. Cogn. Technol. Work **19**(2–3), 477–491 (2017)
3. Duncan, C.J.: Selective attention and the organization of visual information. J. Exper. Psychol. Gen. **113**(4), 501 (1984)
4. Hang, D.: Mission control of multiple unmanned aerial vehicles: a workload analysis. Hum. Factors **47**(3), 479–487 (2005)
5. Cummings, M.L., Bruni, S., Mercier, S., Mitchell, P.J.: Automation architecture for single operator, multiple UAV command and control. Int. C2 J. **1**(2), 1–24 (2007)
6. Pomranky, R.A., Wojciechowski, J.Q.: Determination of mental workload during operation of multiple unmanned systems. Report No. ARL-TR-4309. Army Research Lab Abeerden (2007)
7. Kostenko, A., Rauffet, P., Chauvin, C., Coppin, G.: A dynamic closed-looped and multidimensional model for mental workload evaluation. IFAC-PapersOnLine **49**(19), 549–554 (2016)
8. Wickens, C., Dixon, S., Goh, J., Hammer, B.: Pilot dependence on imperfect diagnostic automation in simulated UAV flights: an attentional visual scanning analysis. In: 13th International Symposium on Aviation Psychology, Dayton, OH (2005)
9. Kostenko, A.: Multidimensional and dynamic evaluation of the control of the situation by the operator: creation of a real-time mental load indicator for drone supervision activity. Thesis dissertation. Université Bretagne Sud, Lorient (2017)
10. Hockey, G.R.J.: Operator functional state: the assessment and prediction of human performance degradation in complex tasks, vol. 355. IOS Press (2003)
11. Zhang, J., Yin, Z., Wang, R.: Recognition of mental workload levels under complex human–machine collaboration by using physiological features and adaptive support vector machines. IEEE Trans. Hum.-Mach. Syst. **45**(2), 200–214 (2015)
12. Yin, Z., Zhang, J.: Operator functional state classification using least-square support vector machine based recursive feature elimination technique. Comput. Methods Programs Biomed. **113**(1), 101–115 (2014)
13. Hollnagel, E.: Context, cognition and control. In: Waern, Y. (ed.) Co-operative Process Management: Cognition and Information Technology, pp. 27–52. Taylor & Francis, London (2003)
14. Parent, M., Gagnon, J.F., Falk, T.H., Tremblay, S.: Modeling the operator functional state for emergency response management. In: ISCRAM (2016)
15. Rauffet, P., Lassalle, J., Leroy, B., Coppin, G., Chauvin, C.: The TAPAS project: facilitating cooperation in hybrid combat air patrols including autonomous UCAVs. Procedia Manuf. **3**, 974–981 (2015)
16. Yang, S., Zhang, J.: An adaptive human–machine control system based on multiple fuzzy predictive models of operator functional state. Biomed. Signal Process. Control **8**(3), 302–310 (2013)
17. Schulte, A., Donath, D., Honecker, F.: Human-system interaction analysis for military pilot activity and mental workload determination. In: 2015 IEEE International Conference on Systems, Man, and Cybernetics (SMC), pp. 1375–1380 (2015)
18. Van Zomeren, A.H., Brouwer, W.H.: Clinical Neuropsychology of Attention. Oxford University Press, New York (1994)

19. Sturm, W., Willmes, K.: On the functional neuroanatomy of intrinsic and phasic alertness. Neuroimage **14**, 76–84 (2001)
20. Posner, M.I.: Measuring alertness. Annal. NY Acad. Sci. **1129**, 193–199 (2008)
21. Valdez, P., Ramírez, C., García, A., Talamantes, J., Armijo, P., Borrani, J.: Circadian rhythms in components of attention. Biol. Rhythm Res. **36**, 57–65 (2005)
22. Rubin, O., Meiran, N.: On the origins of the task mixing cost in the cuing task-switching paradigm. J. Exp. Psychol. Learn. Memory Cogn. **31**, 1477 (2005)
23. Matthews, G., Desmond, P.A.: Task-induced fatigue states and simulated driving performance. Q. J. Exper. Psychol. Sect. A **55**(2), 659–686 (2002)
24. Mulder, G.: The concept and measurement of mental effort. In: Hockey, G.R.J., Gaillard, A. W.K., Coles, M.G.H. (eds.) Energetics and Human Information Processing, pp. 175–198. Springer, Dordrecht (1986). https://doi.org/10.1007/978-94-009-4448-0_12
25. Brown, I.D.: Driver fatigue. Hum. Factors **36**, 298 (1994)
26. Desmond, P.A., Hancock, P.A.: Active and passive fatigue states. In: Hancock, P.A., Desmond, P.A. (eds.) Human Factors in Transportation. Stress, Workload, and Fatigue, pp. 455–465. Lawrence Erlbaum Associates Publishers, Mahwah (2001)
27. Gimeno, T.P., Cerezuela, P.G., Montanes, M.C.: On the concept and measurement of driver drowsiness, fatigue and inattention: implications for countermeasures. Int. J. Veh. Des. **42** (1–2), 67–86 (2006)
28. Zhao, C., Zhao, M., Liu, J., Zheng, C.: Electroencephalogram and electrocardiograph assessment of mental fatigue in a driving simulator. Accid. Anal. Prev. **45**, 83–90 (2012)
29. Lavie, N., Hirst, A., De Fockert, J.W., Viding, E.: Load theory of selective attention and cognitive control. J. Exper. Psychol. Gen. **133**(3), 339 (2004)
30. Bardy, B.G.: Le paradigme de la double tâche. Sci. Motricité **15**, 31–39 (1991)
31. Dehais, F., Fabre, E.F., Roy, R.N.: Cockpit intelligent et interfaces cerveau-machine passives. In: Digital Intelligence (2016)
32. De Waard, D.: The Measurement of Drivers' Mental Workload. Groningen University, Traffic Research Center, Netherlands (1996)
33. Kahneman, D.: Attention and Effort, vol. 1063. Prentice-Hall, Englewood Cliffs (1973)
34. Parasuraman, R.: Neuroergonomics: Research and practice. Theor. Issues Ergon. Sci. **4**(1–2), 5–20 (2003)
35. Strait, M., Scheutz, M.: What we can and cannot (yet) do with functional near infrared spectroscopy. Front. Neurosci. **8**, 117 (2014)
36. Aghajani, H., Garbey, M., Omurtag, A.: Measuring mental workload with EEG+ fNIRS. Front. Hum. Neurosci. **11**, 359 (2017)
37. Verdière, K.J., Roy, R.N., Dehais, F.: Detecting pilot's engagement using fNIRS connectivity features in an automated vs. manual landing scenario. Front. Hum. Neurosci. **12**, 6 (2018)
38. Leplat, J.: La notion de régulation dans l'analyse de l'activité. Perspectives interdisciplinaires sur le travail et la santé, 8–1 (2006)
39. Diaz-Piedra, C., Rieiro, H., Cherino, A., Fuentes, L.J., Catena, A., Di Stasi, L.L.: The effects of flight complexity on gaze entropy: an experimental study with fighter pilots. Appl. Ergon. **77**, 92–99 (2019)
40. Poole, A., Ball, L.J., Phillips, P.: In search of salience: a response-time and eye-movement analysis of bookmark recognition. In: Fincher, S., Markopoulos, P., Moore, D., Ruddle, R. (eds.) People and Computers XVIII—Design for Life, pp. 363–378. Springer, London (2005). https://doi.org/10.1007/1-84628-062-1_23
41. Cegarra, J., Chevalier, A.: The use of Tholos software for combining measures of mental workload: toward theoretical and methodological improvements. Behav. Res. Methods **40** (4), 988–1000 (2008)

42. Durkee, K.T., Pappada, S.M., Ortiz, A.E., Feeney, J.J., Galster, S.M.: System decision framework for augmenting human performance using real-time workload classifiers. In: IEEE International conference on Cognitive Methods in Situation Awareness and Decision Support (CogSIMA) (2015)

43. Sahayadhas, A., Sundaraj, K., Murugappan, M.: Detecting driver drowsiness based on sensors: a review. Sensors 12(12), 16937–16953 (2012)

On the Use of Machine Learning
for EEG-Based Workload Assessment:
Algorithms Comparison in a Realistic Task

Nicolina Sciaraffa[1,2,3(✉)], Pietro Aricò[2,3,4], Gianluca Borghini[2,3,4],
Gianluca Di Flumeri[2,3,4], Antonio Di Florio[2], and Fabio Babiloni[2,4]

[1] Department of Anatomical, Histological, Forensic and Orthopedic Sciences,
Sapienza University of Rome, Rome, Italy
nicolina.sciaraffa@uniroma1.it
[2] BrainSigns srl, Rome, Italy
[3] IRCCS Fondazione Santa Lucia, Rome, Italy
[4] Department of Molecular Medicine, Sapienza University of Rome, Rome, Italy

Abstract. The measurement of the mental workload during real tasks by means
of neurophysiological signals is still challenging. The employment of Machine
Learning techniques has allowed a step forward in this direction, however, most
of the work has dealt with binary classification. This study proposed to examine
the surveys already performed in the context of EEG-based workload classification and to test different machine learning algorithms on real multitasking
activity like the Air Traffic Management. The results obtained on 35 professional Air Traffic Controllers showed that a KNN algorithm allows discriminating up to three workload levels (low, medium and high) with more than 84%
of accuracy on average. Moreover, in such realistic employment it emerges how
important is to opportunely choose the set of features to ward off that task-
related confounds could affect the workload assessment.

Keywords: Workload · EEG · Machine learning · Real settings · ATM

1 Introduction

Reading the Special Issue for the golden anniversary of the "Multiple Resources and
Mental Workload" theory of Christopher D. Wickens [1] allows to retrace the history
of the concept of workload from the difficulty of its definition up to the need to measure
it within Human Factors, passing through the definition of a workload model. In fact,
since the '70s, when the term workload began to appear in scientific publications,
several terms and definitions have overlapped and followed one another. The mental
workload, mental effort and mental strain were the most widely used terms to define the
relationship between the cognitive resources of a person who is required to perform a
task and the difficulty of the task itself. One of the first definitions of workload was "the
mental effort that the human operator devotes to control or supervision relative to his
capacity to expend mental effort" [2]; another typical description define the workload
as "the difference between the capacities of the information processing system that are
required for the task performance to satisfy performance expectations and the capacity

L. Longo and M. C. Leva (Eds.): H-WORKLOAD 2019, CCIS 1107, pp. 170–185, 2019.
https://doi.org/10.1007/978-3-030-32423-0_11

available at any given time" [3]. In each of the definitions given to the workload to date there is the term "capacity" which implies a finite amount of resources [1], in this case cognitive resources. The pool of cognitive resources referred to is not unique, but it is the set of different pools that allows to explain the link between performance and difficulty in the case of multitasking. In fact, many of the actions we perform are multitasking, such as observing a picture to describe it, talking while driving and typically multitasking activities such as those carried out in an aircraft cockpit.

1.1 Multifaceted Aspects of Workload

It is precisely in these safety-critical environments that the need to evaluate an operator's workload was firstly felt. In 1981, again Wickens pointed out that the development of increasingly complex technologies had radically changed the role and load to which an operator was subjected, leading to the dual need to exploit the model of multiple resources to optimize the processing of human operator information in the definition of tasks ("Should one use keyboard or voice? Spoken words, tones, or text? Graphs or digits? Can one ask people to control while engaged in visual search or memory rehearsal?"[1]) and measure the operator's workload. From that moment on, the measure of the workload has spread from the aeronautical [4, 5, 62], the educational [6, 7] and to the clinical [8] fields. Even the aims of workload measurement have evolved: the ultimate goal in all environments is mainly the Workload Adaptation, the process of workload management to aid learning, healing or limiting human errors. Moreover, workload measurement affects both the design and management of interfaces. On the one hand, by testing the workload of subjects during the use of web interfaces [9], for example, it is possible to direct the design. On the other hand, in the field of adaptive automation, it is the continuous monitoring of the workload level of the subject that allows the system to vary the feedback in response to the mental state of the operator [10, 11, 63].

1.2 Workload Measurements

The workload of an operator can be measured in three ways: by administering questionnaires, analyzing the performance of a subject or through psychophysiological measures. Since the workload has different aspects (e.g. mental workload is different from physical workload) the questionnaires preferably used are multidimensional ones, such as NASA TLX [12] and SWAT [13]. However, these are subjective measures and require subjects to be trained in interpreting and judging their condition. Moreover, they can only be assessed after a task, not online. Similarly, performance measures do not represent a direct indicator of the workload status of the subject as they do not allow to know the amount of resources used and therefore the residual resources to reach that performance value [14]. Moreover, measuring performance on a task does not allow to obtain this measure of a differential nature (the remaining resources) so it is always necessary to use a secondary task. However, even the use of secondary tasks very often remains too closely linked to typical laboratory tasks and makes what really happens in multitasking implausible [15]. The main objective of neurophysiological measurements is to provide an objective and continuous, as well as an online measurement of an operator's workload. Thanks to the possibility of making neuro-physiological measures less and less intrusive, so far have been correlated with workload values almost all neurophysiological known measures like the Electrocar-diogram (ECG), the Eye movements, the Pupil diameter, the Respiration, the Galvanic Skin Response (GSR) and the brain activity. Summarizing the evidence, the ECG, the GSR and the ocular activity measurements highlighted a correlation, not only with workload, but also with different mental states like stress, mental fatigue, drowsiness. Therefore, they were demonstrated to be useful and robust only in combination with other neuroimaging techniques directly linked to the Central Nervous System (CNS), that is the brain [16]. Consequently, the electroencephalogram (EEG) and the func-tional Near-Infrared Spectroscopy (fNIRS) as measures of the brain activity, are the most likely candidates that can be straightforwardly employed to monitor the workload in real environments [17]. Between the two, the EEG is usually preferred for the workload assessment for its high temporal resolution. Moreover, it has been proved that EEG features provide higher accuracy respect to ECG and GSR ones [18, 19]. The electroencephalogram is the measure of brain electrical activity that in a non-invasive way can be performed by means of electrodes placed on the scalp. To date, there has been a strong improvement in technology oriented towards minimally invasive sys-tems, with few electrodes, and possibly, dry electrodes [20]. The analysis of EEG signals is usually aimed at studying the variance of the spectral power in the conditions of interest. In the case of Workload, it has emerged that a higher task demand corre-sponds to an increase in frontal theta band activity and a decrease in parietal alpha band activity [16].

1.3   Machine Learning to Get Back Out-of-the-Lab

Therefore, the concept of workload was born for practical needs, has been modeled in the laboratory essentially using dual-task procedures, but then the need to measure it in realistic contexts as in operational, educational and clinical returns overwhelmingly. The practical implications of applying a workload measurement in a realistic envi-ronment define the characteristics that an automatic workload measurement system must have. Firstly, especially in the applications in real environment, it is difficult to create a direct link between the mental state of the subject and his brain activity, or more generally his physiological state since there is no control condition typical of the laboratory. The employment of secondary cognitive task (e.g. the n-back) during real activities does not fit the realistic conditions and may increase the actual workload level [21]. Moreover, because of the high individual variability of physiological responses, traditional statistical tests are not able to discover the relationship between cause and effect, so it is necessary to employ techniques that allow to take into account the individual characteristics to correctly define the level of workload, such the machine learning techniques [22]. Such methods allow to extract the features mostly influenced by the mental state variation, and then use this information to classify the specific workload level. Secondly, since by definition the workload is linked to the performance by the inverted "U" model [23], the ideal would be to be able to distinguish at least 3 levels of workload, one suboptimal that concerns the workload too low, one optimal, and the threshold that defines the overload condition. However, most of the work in literature is limited to classify two levels of workload, the low and high. In these cases, the levels of accuracy reached are generally very high, greater than 80% [18, 22, 24–29]. Much less are the examples of multiclass-classification [17], whose highest number of workload levels classified has been 7 [30] and, almost all, have been obtained by means of n-back and arithmetic tasks in a laboratory context [31, 32]. In this context, the majority of methods used to define the level of mental workload of a subject are supervised machine learning techniques. The process that leads from the recording of EEG signals to an indication of the workload level passes through the use of signal analysis methods that allow to extract the informative features of the phenomenon to be investigated. Regarding the measurement of the workload have been used in several studies both spectral, temporal and spatial features [33]. The use of spectral features remains the most suitable for the temporal continuity required by the workload monitoring, since the brain activity induces variations in its spectral power which, unlike ERPs used for time domain analysis [34], does not need to be triggered with a certain timing [35]. Taking into account the nature of the features, there are countless different examples of configurations, in terms of number of channels and frequencies used in the literature. The number of electrodes can vary from 64 [24, 36] to 6 [37], and even the bands considered vary from 2 (Theta and Alpha, [9]), to 7 (0–4 Hz, 4–7 Hz, 7–12 Hz, 12–30 Hz, 30–42 Hz, 42–84 Hz, 84–128 Hz [38]), up to considering all the single frequency bins that define the spectrum [39]. Several studies have shown that it does not necessarily take more than 5–10 electrodes to classify the workload [24]. Especially when dealing with a high number of channels and high spectral resolution, the number of used features increases exponentially and leads to the so-called "curse of dimensionality"[40]. Many researchers have therefore highlighted the need to make a selection of features both to decrease the computational cost of machine learning algorithms and to use this as additional information in experimental setups. In fact, if the analysis shows that some electrodes are not useful for the classification of the workload, it is possible to remove them and then make the instrumentation lighter and less invasive. In this case the most used methods for the selection of features are those recursive, such as recursive feature elimination [18, 41], sequential forward feature selection [24] or methods that take into account the dependence between features such as the Minimum Redundancy Maximum Relevance selection [37, 42, 43], or unsupervised method (Locally linear embedding, [44]). Once defined a meaningful set of features, it is necessary to choose a model to define the workload level. In the literature innumerable algorithms, essentially of a supervised nature, have been used to define the workload level of a subject starting from his brain signals, belonging to both the so-called Shallow learning and deep Learning domains [45, 46]. In all cases the efficiency of such algorithms is usually presented in terms of accuracy. However, the accuracy obtained in different studies are not directly comparable, since the calculation of accuracy includes several factors like the task employed to elicit the workload, the number of subjects and the number of EEG signals recorded (and also the kind of instrumentation employed), the features extracted, the methods used to eventually select them and, only at the end, the algorithm used to classify the workload. Only taking into account all this information, it could be possible to compare the results obtained so far in different works employing machine learning techniques to classify the level of workload. Leveraging on the theoretical comparison of the works done so far with regard to the classification of the workload through electroencephalographic signals, it is necessary to highlight an issue. In any case, starting from a very large set of features or making a blind selection of them, very high accuracy of classification could be obtained. However, it could not be directly associated with a change in the workload level. To be sure that the system has actually classified workload, it is therefore necessary to perform two fundamental actions. Firstly, it is necessary to perform an excellent calibration of the machine learning algorithm avoiding task-related confound, like for example movements or the influences of other mental states. However, during a real task that is typically highly multitasking, it may not be possible to perform a rigorous calibration of the system, as the ideal conditions provided by the typical control conditions of the laboratory task are lacking. Therefore, in these cases the calibration could be dirtied by task-related confounds. To solve this problem cross-task calibration has been proposed, but the results produced so far have not shown that it is possible to use a laboratory task to effectively classify a real task and the performance is much lower than that obtained in a within-subject condition [41]. Secondly, it should be preferred a careful selection of features related to the phenomenon, and possibly not to the other mental state variations, respect to a blind one. In the case of workload, for example, many works have identified in the activity of the frontal brain areas in the theta band and parietal areas in the alpha band [16] the most informative features. Even more accurately, [30] have carried out a selection of the channels through source localization analysis, and it has been possible for them to classify 7 levels of workload. Therefore, the practical aim of this work is to provide a comparison of five different machine learning algorithms and two different sets of EEG features to discriminate three different levels of workload during a real task of Air Traffic Management. The Air Traffic Management (ATM) is a highly multitasking activity. In fact, air traffic controllers are continuously engaged in visual activities (airplane control on the radar) and auditory (as they communicate with the pilots of different aircrafts). The ATM represents one of the fields where the evaluation of workload has a fundamental role both in training aspects and in operational conditions [47, 48]. In fact, it has been established that it is a high demanding work during which the slightest error could have very ominous effects [49]. Changes in the traffic manipulation, complexity or volume, produced changes in mental resources required and therefore in workload. In ATM domain, different tasks have been used to investigate workload changes and to create a workload index based on biosignals, even though most of the results are related to laboratory environment. However, the present study takes advantages of realistic task in a highly realistic context and of 35 professional Air Traffic Controllers.

2   Methods

2.1   Experimental Protocol

An experimental protocol with high realistic ATM scenarios has been settled up. The controller position is similar to the ones used in the real operational center. The controller working position has two screens, one 30″ to display radar image and a 21″ to interact with the radar image (zoom, move, clearances and information) and the voice communication between controller and pseudo-pilot uses the same hardware and software like the one used in training (headphones, microphone and push-to-talk command), very similar to what normally used into operations. The radar picture shows the sector (light grey), routes, waypoints and flights displayed according to their coordination state (white ones are assumed). Information on neighbor flights is displayed in the list. The experimental task consists in an ATM scenario in which different air-traffic conditions take place. For instance, the task could start with an increasing traffic complexity until a hard condition, and then decreasing until a condition similar to the initial one by passing through a medium complexity condition. The variation of the task complexity is necessary to evaluate if the system is able to differentiate the different workload levels. Each scenario lasts globally 45 min, while each session of low (L), medium (M) and high (H) workload level lasts about 15 min. The different levels of traffic, defined by subject matter experts (SME), vary according to the number of aircrafts, traffic geometry, the number of conflicts and subjective assessment of controller's skill. Three different scenarios have been proposed, with compatible events in order to induce the three mentioned difficulty levels (Easy, Medium, Hard), but not exactly the same, to not induce habituation or expectation effects on the experimental subjects. The experimental protocol involves 35 professional ATCOs. ATCOs have been selected in order to have a homogeneous sample in terms of sex, age and expertise. Sixteen EEG channels (FPz, F3, Fz, F4, AF3, AF4, C3, Cz, C4, P3, Pz, P4, POz, O1, Oz, O2) have been recorded by the digital monitoring BEmicro system (EBNeuro system) with a sampling frequency of 256 Hz. All the EEG electrodes have been referenced to both earlobes, and the impedances of the electrodes have been kept below 10 kΩ.

2.2 Signal Processing

The EEG signals have been digitally band-pass filtered by a 4th order Butterworth filter (low-pass filter cutoff frequency: 30 Hz, high-pass filter cutoff frequency: 1 Hz) and the Fpz signal has been used to correct eyes-blink artifacts from the EEG data by means of the Reblinca algorithm [50]. It should be underlined that normally in a realistic environment, different sources of artifacts could affect neurophysiological recorded signals, more than in the laboratory environments. For instance, ATCOs normally communicate verbally and perform several movements during their operational activity. Then each trial having an amplitude exceeds 100 µV (threshold criteria), or the slope trend higher than 3, or a sample to sample difference higher than 25 µV have been marked as "artifact" and then rejected, with the aim to have clean EEG signals from which estimate the brain parameters for the different analyses. The aforementioned parameters and related techniques have been set following the methodology available on the EEGLAB toolbox [51].

2.3 Features Extraction

The EEG signals have been segmented in periods of 2 s, 0.125 s shifted. After that, for each period, the power spectral density (PSD) by using the Fast Fourier Transform has been computed in the Theta and Alpha frequency bands because it has been stated that they are the most related to workload effects [9, 16]. The EEG frequency bands were defined accordingly with the Individual Alpha Frequency (IAF) value estimated for each subject. Since the alpha peak is mainly prominent during rest conditions, the participants were asked to keep the eyes closed for a minute before starting with the experiment. In particular, the theta and alpha bands have been respectively defined as (IAF−6 ÷ IAF−2) and (IAF−2 ÷ IAF+2). Two different sets of features have been considered. In the first case, the PSD values in the theta and alpha bands have been computed for 14 EEG electrodes, to imitate what is usually done in several studies in literature [9, 41, 43, 52]. In this work, 28 PSD Features (14 Channels × 2 Bands) have been computed. The second set consisted of 9 features, 5 describing the frontal theta activity and 4 the alpha parietal activity. These features have been chosen according to the literature, because it is proven these are the features most correlated to workload [6, 16]. In both cases the features have been normalized because the differences in ranges affect the calibration and the functioning of some algorithms [53] (e.g. K nearest neighbor).

2.4   Machine Learning Algorithms

Five different machine learning techniques have been employed to discriminate between three different levels of workload (i.e. Low, Medium and High level). The starting dataset shows balanced classes (6000 instances per class) and has been used in a within-subject manner, as this approach is allowed in case of long-lasting recordings of a single subject. The total amount of occurrences available for each subject has been divided in an optimization set, which occurrences have been used for the optimization of model parameters by using grid search and 3-fold cross validation, and an evaluation set, where the performance of the algorithms have been evaluated computing the accuracy through 5-fold cross-validation. Optimization and evaluation of machine-learning techniques have been computed by Phyton Scikit-learn library [54]. In particular five different techniques have been trained to cover a wide range of algorithms types: a regression-based method (the multinomial logistic regression), a linear method without any optimization procedure (LDA), a linear classifier with a cost parameter (SVM), an instance-based method (the k nearest neighbor) and an ensemble method (the Random Forest).

Logistic Regression (LR) is a model used for the prediction of the probability of occurrence of an event. In this case it has been used in its multiclass configuration, the multinomial logistic regression and the value of l_2 penalization has been chosen in the log space between −3 and 3

Linear Discriminant Analysis (LDA) is a linear algorithm that allows to create hyperplanes in n-dimensional space according to the number of features, to discriminate 2 or more classes. Its advantage is that it has not any parameter to optimize, however, it could finally try to describe only linear problems.

Support Vector Machine (SVM) is a supervised algorithm that allows to create hyperplanes in n-dimensional space according to the number of features, to discriminate 2 or more classes. It could be a linear or nonlinear classifier (or regressor) according to the employed kernel. In this case it has been used a linear kernel in a multiclass configuration using the Crammer and Singer method () and the optimal cost parameter has been chosen in a log space between −3 and 3.

K-Nearest Neighbor (KNN) is a nonlinear instance-based algorithm. Its main idea is to predict the class on the basis of the distance between the observation and the first k neighbors and does not assume a priori the distribution of the dataset. The advantage of this algorithm is that it is optimized locally, and it is not affected by the complexity of the entire phenomena. The weakness is that the computational cost could be as high as the amount of features increase. Only the number k of neighbors has been chosen in a range from 1 to 20.

Random Forest Classifier (RF) is a nonlinear classifier [55] belonging to the ensemble methods. This family of classifiers allows to generalize well to new data [56] and are more robust to overfitting than individual trees because each node does not see all the features at the same time [55]. Several parameters could be optimized, however, in this case only the number of trees ([10, 100, 200]) have been chosen.

In the latter case, the algorithm allows also to obtain the information related to the feature importance that could be used to explain how the model is affected by each feature. Therefore the topographies showing the feature importance in the case of 28-feature and 9-feature sets have been compared.

3 Results

The classification results for both sets of features are shown in Figs. 1 and 2. Each box plot represents the value of the accuracy of the population, while the single mean value of the accuracy obtained for each subject is shown in the black dot. The Friedman test has been performed to statistically assess if there is any significant difference between the algorithms, because the sphericity requirement was not met for both conditions (Mauchly's test $p < 10^{-4}$). For the 28-feature set the Friedman test provided χ^2 (N = 35, df = 4) = 123.0171 ($p < 10^{-4}$) and for the 9-feature set χ^2 (N = 35, df = 4) = 126.2857 ($p < 10^{-4}$). The multiple comparison Bonferroni corrected has been performed and the results are shown in Table 1. In both cases the mean accuracy provided by the KNN is significantly higher than all the other algorithms. The Random Forest provided significantly higher accuracy than the LR, LDA and SVM. Such 3 methods did not show significant differences in their performances when the 28-feature set was used, while the SVM performed significantly worse than the other algorithms when 9 features were used. In Fig. 3 are shown the topographies of the feature importance computed with the Random Forest algorithm for theta and alpha band. The values have been normalized. The scalp maps show that the most important features are the central and occipital PSD values in alpha band in the case of 28-feature set and the parietal activity in alpha band in case of the 9-feature set.

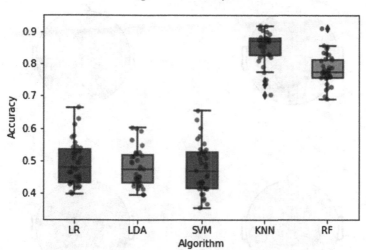

Fig. 1. Accuracy distribution for each algorithm in case of 28 features. The black dots represent the accuracy value for each subject.

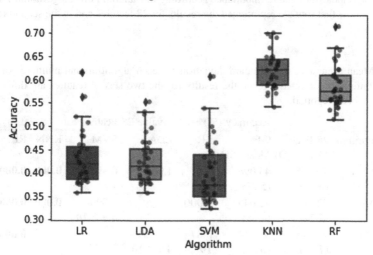

Fig. 2. Accuracy distribution for each algorithm for 9-feature set. The black dots represent the accuracy value for each subject.

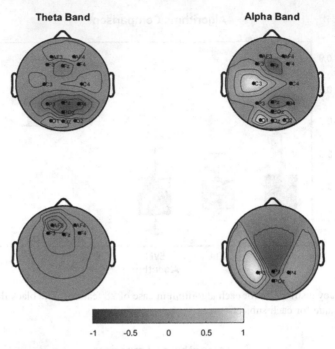

Fig. 3. Topographies of Feature importance according to Random Forest algorithm in theta and alpha band. In the first row are represented the results for 28 features and in the second row for 9 features.

Table 1. Mean accuracies and standard deviation for each algorithm and multiple comparison p-value (Bonferroni corrected). When the results for the two sets of features are different both p-value have been reported.

Algorithm	Accuracy (%) 28-Feat Mean (SD)	9-Feat Mean (SD)	p-value 28-Feat/9 Feat LR	LDA	SVM	KNN	RF
LR	49.176 (6.74)	43.096 (5.72)		1.000	0.874/ 10^{-7}	**0.000**	**0.000**
LDA	47.889 (5.74)	42.645 (4.88)	1.000		1.000/ $4.9 * 10^{-5}$	**0.000**	**0.000**
SVM	47.645 (7.57)	39.914 (6.55)	0.874/ 10^{-7}	1.000/ $4.9 * 10^{-5}$			**0.000**
KNN	84.947 (5.05)	62.112 (3.98)	**0.000**	**0.000**	**0.000**	**0.000**	**0.000**
RF	78.214 (4.68)	58.617 (4.57)	**0.000**	**0.000**	**0.000**	**0.000**	

4 Discussion

This work aims to classify three different levels of workload during real multitasking activities like the ATM. According to the theory, a correct modulation of the workload in a laboratory environment should be based on a dual task to allow an evaluation of the subject's residual cognitive resources and consequently, by definition, of his workload. However, in a real environment, it is very difficult to integrate a control condition, as well as to take into account the variability underlying cognitive phenomena both intra and inter subject. Therefore, the application of Machine Learning has been considered the solution for classifying the workload and overcoming these issues typical of real applications. The preliminary analysis of the works carried out so far in this context has shown that it is possible to discriminate with acceptable accuracy only two levels of workload [6, 18, 22, 24, 25, 27–29, 33, 36–39, 42, 45, 46, 57], even though, above all in view of a practical application of the workload measurement, it is necessary to establish at least the value of two thresholds to define the underload and the overload state. The most frequently employed features are the spectral ones, because they can be calculated with a high temporal resolution (up to one second) and allow to monitor brain activity in a quantitative manner without temporal triggers. Therefore, in this work the values of the PSD calculated in time windows of 2 s have been used, averaging the values of PSD in each band to limit the number of features and keep at the same time under control the collinearity [58]. Due to this condition, in fact, the information provided by very close frequency bins could be superimposable, which would lead to introduce a bias in the creation of the model. Since the number of observations available is in the order of thousands, it was decided not to use any feature selection algorithm, but to provide a posteriori information on the feature importance. One of the chosen algorithms, the Random Forest, allowed to have this information. Taking advantage of this potentiality, it has been highlighted that the most discriminating features of the concerned model are in the alpha band. In particular, when the higher number of features has been used, the most important features cover the central and occipital brain areas. This aspect can be explained considering that the alpha band intervenes twice in the considered task. In fact, in the alpha band it is possible to find both the motor alpha pattern, due to the activity of the sensorimotor area and generally strongly lateralized [59], and the pattern associated to the visual area. This set of features is not directly referring to the typical workload topographies, whose purpose is to measure the net of cognitive resources used by the subject, but rather these features provide the information derived from the movements of the subjects to define the level of workload imposed by the task itself, which does not necessarily correspond to that perceived by the subject. On the other hand, when only frontal theta and parietal alpha features have been used, the most important features are related to parietal activity, that is usually associated with the attentional alpha pattern that reflects the increase of brain activity in areas afferent to the posterior attentional networks [59]. Therefore, it has been demonstrated that, especially when the task is real and the algorithm does not take advantage of a rigorous calibration to avoid the task related confounds, the role of the features chosen a priori becomes essential and recalls the concept of "no free lunch" [60] in machine learning: the necessity to use prior knowledge to optimize the algorithm functioning, but at the cost of generality. However, it is necessary to highlight that reducing the number of features there is a decreasing in the performance of each tested algorithms. In fact, to classify three levels of workload it was decided to test different algorithms of machine learning, which in the first place can be divided into two categories: regression and classification. Among the classifiers, we can further distinguish linear (LDA and SVM) and non-linear (KNN and RF) methods. Although regression has been proposed as a method that avoiding the strict equality between classes allowing to have higher performances, especially in case of cross-task classification [41], it provided an accuracy value equal to 50%, but still higher than the chance (33.33%). On the other hand, linear methods both in the case of optimization (SVM) and in the case of non-optimized method (LDA) provided significantly lower accuracy than nonlinear methods. The linear methods are appropriate when limited data and limited knowledge about the data itself is available [61]. However, a linear classifier does not work well in the presence of strong noise or outliers, if the dimensionality of the features space is too high, if regularization was not done well or the problem is intrinsically non-linear [17]. If there are large amounts of data, non-linear methods are suitable to find potentially complex structure in the data. In this work the KNN not only provides the highest accuracy (84%), but it also has different advantages: it is a method that does not require the calculation of the covariance matrix as in the case of the LDA and is therefore mathematically very simple [56]. On the other hand, it does not need time for training (because it just memorizes the training set) and then could be used for an online application when there are a few features. In fact, in the case of a large number of features this method does not allow to easily manage those irrelevant and at the same time becomes computationally expensive to calculate the distance. In addition to the high accuracy provided, even the choice of Random Forest as classifier in realistic multitasking could be advantageous essentially for two reasons. First, because it is an ensemble method, it tends to generalize well and is not subjected to overfitting, Second, it allows to have the information regarding the feature importance, that increase the possibility to know what the system is actually classifying. However, the final choice of a classifier should be made after a systematic evaluation of other different performance parameters, such sensitivity, specificity, recall and precision.

5 Conclusion

With this work it has been proved that it is possible to reach very high accuracy to distinguish between three levels of workload during a real task only by using the EEG signals. However, according to the literature the high accuracy is only one of the optimal characteristics required for an out-of-the-lab classifier besides the none or at most few data samples for training the classifier and higher temporal reliability. Therefore, several other questions need to be pointed out in realistic contexts.

Acknowledgment. This work is co-financed by the European Commission by Horizon2020 projects "WORKINGAGE: Smart Working environments for all Ages" (GA n. 826232); "SIMUSAFE": Simulator Of Behavioural Aspects For Safer Transport (GA n. 723386);

"SAFEMODE" (GA n. 814961); "BRAINSAFEDRIVE: A Technology to detect Mental States during Drive for improving the Safety of the road" (Italy-Sweden collaboration) with a grant of Ministero dell'Istruzione dell'Università e della Ricerca della Repubblica Italiana.

Bibliography

1. Wickens, C.D.: Multiple resources and mental workload. Hum. Factors **50**(3), 449–455 (2008)
2. Curry, R., Jex, H., Levison, W., Stassen, H.: Final report of control engineering group. In: Moray, N. (ed.) Mental Workload. NATO Conference Series, vol. 8, pp. 235–252. Springer, Boston (1979). https://doi.org/10.1007/978-1-4757-0884-4_13
3. Gopher, D.: In defence of resources: on structures, energies, pools and the allocation of attention. In: Hockey, G.R.J., Gaillard, A.W.K., Coles, M.G.H. (eds.) Energetics and Human Information Processing. NATO ASI Series (Series D: Behavioural and Social Sciences), vol. 31, pp. 353–371. Springer, Dordrecht (1986). https://doi.org/10.1007/978-94-009-4448-0_25
4. Kantowitz, B.H., Casper, P.A.: Human workload in aviation. In: Human Error in Aviation, pp. 123–153. Routledge, Abingdon (2017)
5. Bargiotas, I., Nicolaï, A., Vidal, P.-P., Labourdette, C., Vayatis, N., Buffat, S.: The complementary role of activity context in the mental workload evaluation of helicopter pilots: a multi-tasking learning approach. In: Longo, L., Leva, M.C. (eds.) H-WORKLOAD 2018. CCIS, vol. 1012, pp. 222–238. Springer, Cham (2019). https://doi.org/10.1007/978-3-030-14273-5_13
6. Gerjets, P., Walter, C., Rosenstiel, W., Bogdan, M., Zander, T.O.: Cognitive state monitoring and the design of adaptive instruction in digital environments: lessons learned from cognitive workload assessment using a passive brain-computer interface approach. Front. Neurosci. **8**(DEC), 1–21 (2014)
7. Byrne, A.: The effect of education and training on mental workload in medical education. In: Longo, L., Leva, M.C. (eds.) H-WORKLOAD 2018. CCIS, vol. 1012, pp. 258–266. Springer, Cham (2019). https://doi.org/10.1007/978-3-030-14273-5_15
8. Byrne, A.: Measurement of mental workload in clinical medicine: a review study. Anesthesiol. Pain Med. **1**(2), 90 (2011)
9. Jimenez-Molina, A., Retamal, C., Lira, H.: Using psychophysiological sensors to assess mental workload during web browsing. Sensors (Switzerland) **18**(2), 1–26 (2018)
10. Aricò, P., Borghini, G., Di Flumeri, G., Colosimo, A., Pozzi, S., Babiloni, F.: A passive brain–computer interface application for the mental workload assessment on professional air traffic controllers during realistic air traffic control tasks. Prog. Brain Res. **228**, 295–328 (2016)
11. Aricò, P., et al.: Adaptive automation triggered by EEG-based mental workload index: a passive brain-computer interface application in realistic air traffic control environment. Front. Hum. Neurosci. **10**, 539 (2016)
12. Hart, S.G., Staveland, L.E.: Development of NASA-TLX (Task Load Index): results of empirical and theoretical research. Advances in psychology **52**, 139–183 (1988)
13. Reid, G.B., Nygren, T.E.: The subjective workload assessment technique: a scaling procedure for measuring mental workload. Advances in psychology **52**, 185–218 (1988)
14. Wilson, G.F.: Operator functional state assessment for adaptive automation implementation. In: Biomonitoring for Physiological and Cognitive Performance during Military Operations, vol. 5797, pp. 100–105 (2005)

15. Colle, H.A., Reid, G.B.: Double trade-off curves with different cognitive processing combinations: testing the cancellation axiom of mental workload measurement theory. Hum. Factors 41(1), 35–50 (1999)
16. Borghini, G., Astolfi, L., Vecchiato, G., Mattia, D., Babiloni, F.: Measuring neurophysiological signals in aircraft pilots and car drivers for the assessment of mental workload, fatigue and drowsiness. Neurosci. Biobehav. Rev. 44, 58–75 (2014)
17. Aricò, P., Borghini, G., Di Flumeri, G., Sciaraffa, N., Colosimo, A., Babiloni, F.: Passive BCI in operational environments: insights, recent advances, and future trends. IEEE Trans. Biomed. Eng. 64(7), 1431–1436 (2017)
18. Zhang, H., Zhu, Y., Maniyeri, J., Guan, C.: Detection of variations in cognitive workload using multi-modality physiological sensors and a large margin unbiased regression machine. In: 2014 36th Annual International Conference of the IEEE Engineering Medicine and Biology Society EMBC 2014, pp. 2985–2988 (2014)
19. Aricò, P., et al.: Towards a multimodal bioelectrical framework for the online mental workload evaluation. In: 2014 36th Annual International Conference of the IEEE Engineering in Medicine and Biology Society (EMBC), pp. 3001–3004 (2014)
20. Di Flumeri, G., Aricò, P., Borghini, G., Sciaraffa, N., Di Florio, A., Babiloni, F.: The dry revolution: evaluation of three different eeg dry electrode types in terms of signal spectral features, mental states classification and usability. Sensors 19(6), 1365 (2019)
21. Heard, J., Harriott, C.E., Adams, J.A.: A survey of workload assessment algorithms. IEEE Trans. Hum. Mach. Syst. 48(5), 434–451 (2018)
22. Dijksterhuis, C., De Waard, D., Brookhuis, K.A., Mulder, B.L.J.M., De Jong, R., Kerick, S. E.: Classifying visuomotor workload in a driving simulator using subject specific spatial brain patterns. Front. Neurosci. 7(August), 1–11 (2013)
23. Bruggen, A.: An empirical investigation of the relationship between workload and performance. Manag. Decis. 53(10), 2377–2389 (2015)
24. Mathan, S., Smart, A., Ververs, T., Feuerstein, M.: Towards an index of cognitive efficacy: EEG-based estimation of cognitive load among individuals experiencing cancerrelated cognitive decline. In: 2010 Annual International Conference of the IEEE Engineering in Medicine and Biology Society, EMBC 2010, pp. 6595–6598 (2010)
25. Baldwin, C.L., Penaranda, B.N.: Adaptive training using an artificial neural network and EEG metrics for within- and cross-task workload classification. Neuroimage 59(1), 48–56 (2012)
26. Wang, Y.-T., et al.: Developing an EEG-based on-line closed-loop lapse detection and mitigation system. Front. Neurosci. 8, 321 (2014)
27. De Massari, D., et al.: Fast mental states decoding in mixed reality. Front. Behav. Neurosci. 8(November), 1–9 (2014)
28. Schultze-Kraft, M., Dähne, S., Gugler, M., Curio, G., Blankertz, B.: Unsupervised classification of operator workload from brain signals. J. Neural Eng. 13(3), 36008 (2016)
29. Dimitrakopoulos, G.N., et al.: Task-independent mental workload classification based upon common multiband eeg cortical connectivity. IEEE Trans. Neural Syst. Rehabil. Eng. 25(11), 1940–1949 (2017)
30. Zarjam, P., Epps, J., Chen, F., Lovell, N.H.: Estimating cognitive workload using wavelet entropy-based features during an arithmetic task. Comput. Biol. Med. 43(12), 2186–2195 (2013)
31. Aghajani, H., Garbey, M., Omurtag, A.: Measuring mental workload with EEG+fNIRS. Front. Hum. Neurosci. 11(July), 1–20 (2017)
32. Rebsamen, B., Kwok, K., Penney, T.B.: Evaluation of cognitive workload from EEG during a mental arithmetic task. Proc. Hum. Factors Ergon. Soc. 5, 1342–1345 (2011)

33. Zhang, P., Wang, X., Zhang, W., Chen, J.: Learning spatial-spectral-temporal EEG features with recurrent 3D convolutional neural networks for cross-task mental workload assessment. IEEE Trans. Neural Syst. Rehabil. Eng. **27**(1), 31–42 (2019)
34. Aricò, P., Aloise, F., Schettini, F., Salinari, S., Mattia, D., Cincotti, F.: Influence of P300 latency jitter on event related potential-based brain–computer interface performance. J. Neural Eng. **11**(3), 35008 (2014)
35. Radüntz, T.: Dual frequency head maps: a new method for indexing mental workload continuously during execution of cognitive tasks. Front. Physiol. **8**(DEC), 1–15 (2017)
36. Jao, P.K., Chavarriaga, R., Millan, J.D.R.: Using robust principal component analysis to reduce EEG intra-trial variability. In: Proceedings Annual International Conference of the IEEE Engineering Medicine and Biology Society EMBS, vol. 2018-July, no. 1, pp. 1956–1959 (2018)
37. Dehais, F., et al.: Monitoring pilot's mental workload using ERPs and spectral power with a six-dry-electrode EEG system in real flight conditions. Sensors (Switzerland) **19**(6), 1324 (2019)
38. Casson, A.J.: Artificial Neural Network classification of operator workload with an assessment of time variation and noise-enhancement to increase performance, vol. 8, no. December, pp. 1–10 (2014)
39. Fan, J., et al.: A step towards EEG-based brain computer interface for autism intervention. In: 2015 37th Annual International Conference of the IEEE Engineering in Medicine and Biology Society (EMBC), pp. 3767–3770 (2015)
40. Bellman, R., Kalaba, R.: Dynamic programming and statistical communication theory. Proc. Natl. Acad. Sci. U. S. A. **43**(8), 749 (1957)
41. Ke, Y., et al.: An EEG-based mental workload estimator trained on working memory task can work well under simulated multi-attribute task. Front. Hum. Neurosci. **8**(September), 1–10 (2014)
42. Mühl, C., Jeunet, C., Lotte, F., Hogervorst, M.A.: EEG-based workload estimation across affective contexts. Front. Neurosci. **8**(June), 1–15 (2014)
43. Arvaneh, M., Umilta, A., Robertson, I.H.: Filter bank common spatial patterns in mental workload estimation. In: Proceedings Annual International Conference of the IEEE Engineering Medicine and Biology Society EMBS, 2015 November, pp. 4749–4752 (2015)
44. Yin, Z., Zhang, J.: Identification of temporal variations in mental workload using locally-linear-embedding-based EEG feature reduction and support-vector-machine-based clustering and classification techniques. Comput. Methods Programs Biomed. **115**(3), 119–134 (2014)
45. Christensen, J.C., Estepp, J.R., Wilson, G.F., Russell, C.A.: The effects of day-to-day variability of physiological data on operator functional state classification. Neuroimage **59**(1), 57–63 (2012)
46. Yang, S., Yin, Z., Wang, Y., Zhang, W., Wang, Y., Zhang, J.: Assessing cognitive mental workload via EEG signals and an ensemble deep learning classifier based on denoising autoencoders. Comput. Biol. Med. **109**(April), 159–170 (2019)
47. Radüntz, T., Fürstenau, N., Tews, A., Rabe, L., Meffert, B.: The effect of an exceptional event on the subjectively experienced workload of air traffic controllers. In: Longo, L., Leva, M.C. (eds.) H-WORKLOAD 2018. CCIS, vol. 1012, pp. 239–257. Springer, Cham (2019). https://doi.org/10.1007/978-3-030-14273-5_14
48. Edwards, T., Martin, L., Bienert, N., Mercer, J.: The relationship between workload and performance in air traffic control: exploring the influence of levels of automation and variation in task demand. In: Longo, L., Leva, M.C. (eds.) H-WORKLOAD 2017. CCIS, vol. 726, pp. 120–139. Springer, Cham (2017). https://doi.org/10.1007/978-3-319-61061-0_8

49. Arico, P., et al.: Human factors and neurophysiological metrics in air traffic control: a critical review. IEEE Rev. Biomed. Eng. **10**, 250–263 (2017)

50. Di Flumeri, G., Aricò, P., Borghini, G., Colosimo, A., Babiloni, F.: A new regression-based method for the eye blinks artifacts correction in the EEG signal, without using any EOG channel, vol. 2016 (2016)

51. Delorme, A., Makeig, S.: EEGLAB: an open source toolbox for analysis of single-trial EEG dynamics including independent component analysis. J. Neurosci. Methods **134**(1), 9–21 (2004)

52. Lim, W.L., Sourina, O., Wang, L.P.: STEW: simultaneous task EEG workload data set. IEEE Trans. Neural Syst. Rehabil. Eng. **26**(11), 2106–2114 (2018)

53. Nilsson, N.J., Nilsson, N.J.: Artificial Intelligence: A New Synthesis. Morgan Kaufmann, Burlington (1998)

54. Pedregosa, F., et al.: Scikit-learn: machine learning in Python. J. Mach. Learn. Res. **12**(Oct), 2825–2830 (2011)

55. Breiman, L.: Random forests. Mach. Learn. **45**(1), 5–32 (2001)

56. Novak, D., Mihelj, M., Munih, M.: A survey of methods for data fusion and system adaptation using autonomic nervous system responses in physiological computing. Interact. Comput. **24**(3), 154–172 (2012)

57. Hernández, L.G., Mozos, O.M., Ferrández, J.M., Antelis, J.M.: EEG-based detection of braking intention under different car driving conditions. Front. Neuroinformatics **12**(May), 1–14 (2018)

58. Næs, T., Mevik, B.: Understanding the collinearity problem in regression and discriminant analysis. J. Chemom. J. Chemom. Soc. **15**(4), 413–426 (2001)

59. Deiber, M.-P., Sallard, E., Ludwig, C., Ghezzi, C., Barral, J., Ibañez, V.: EEG alpha activity reflects motor preparation rather than the mode of action selection. Front. Integr. Neurosci. **6**, 59 (2012)

60. Wolpert, D.H., Macready, W.G.: No free lunch theorems for search. Technical report SFI-TR-95-02-010, Santa Fe Institute (1995)

61. Wolpaw, J.R., et al.: Brain-computer interface technology: a review of the first international meeting. IEEE Trans. Rehabil. Eng. **8**(2), 164–173 (2000)

62. Di Flumeri, G., et al.: On the use of cognitive neurometric indexes in aeronautic and air traffic management environments. In: Blankertz, B., Jacucci, G., Gamberini, L., Spagnolli, A., Freeman, J. (eds) Symbiotic Interaction. Symbiotic 2015. LNCS, vol. 9359. Springer, Cham (2015). https://doi.org/10.1007/978-3-319-24917-9_5

63. Aricò, P., et al.: Human-machine interaction assessment by neurophysiological measures: a study on professional air traffic controllers. In: 40th Annual International Conference of the IEEE Engineering in Medicine and Biology Society (EMBC) (2018)

Do Cultural Differences Play a Role in the Relationship Between Time Pressure, Workload and Student Well-Being?

Omolaso Omosehin$^{(\boxtimes)}$ and Andrew P. Smith

Centre for Occupational and Health Psychology, School of Psychology,
Cardiff University, 63 Park Place, Cardiff CF10 3AS, UK
omosehino@cardiff.ac.uk

Abstract. Student workload is an issue that has implications for undergraduate student learning, achievement and well-being. Time pressure, although not the only factor that influences students' workload or their perception of it, is very pivotal to students' workload. This may vary from one country to the other and maybe affected by cultural differences. The current study investigated the impact of nationality and time pressure on well-being outcomes as well as perceptions of academic stress and academic work efficiency. The study was cross-cultural and cross-sectional in nature and comprised 360 university undergraduates from three distinct cultural backgrounds: White British, Ethnic Minorities (in the United Kingdom) and Nigerian. The findings suggest that time pressure directly or indirectly (i.e. in tandem with nationality) predicted negative outcomes, work efficiency and academic stress. This implies that nationality/ethnicity also plays a role in the process.

Keywords: Time pressure · Student workload · Cultural differences ·
Nationality · Ethnicity · Student Well-being

1 Introduction

1.1 Conceptualizing Student Workload

Human mental workload is probably the most researched concept "in human factors research and practice" [1]. It describes the demands mental tasks place on the information processing skills of the brain in a way that is analogous to how physical demands tax the muscles [1]. It, thus, can be deduced that workload in occupational settings hinges heavily on occupational demands [2]. Fan and Smith [3] assert that "job demands refer to workload". In student settings, however, workload has been defined as "the time needed for contact and independent study, the quantity and level of difficulty of the work, and the type and timing of assessments, the institutional factors such as teaching and resources, and student characteristics such as ability, motivation and effort" [4]. It can be deduced from this definition that time plays a crucial role in student workload. However, it has been argued that time is not the only factor that contributes to workload [4–7]. As this definition shows, there are other factors that affect workload. Some of these factors as listed by Bowyer [4] include "institutional

© Springer Nature Switzerland AG 2019
L. Longo and M. C. Leva (Eds.): H-WORKLOAD 2019, CCIS 1107, pp. 186–204, 2019.
https://doi.org/10.1007/978-3-030-32423-0_12

factors" like the quality and method of teaching, how much support staff are willing to give students within and beyond the classroom etc; difficulty of the subject matter; how individual modules are planned in terms of assessment submission dates; type of assessment; allocation of time between "contact time" and personal study; and the student's ability, motivation and effort. High workload has been associated with the tendency for students to adopt surface learning strategies as opposed to deep learning approaches [8, 9], and an increased likelihood to cheat and plagiarize [8] as well as the intention to quit university [4]. Student workload is divided into objective and sub-jective workload [4]. While objective workload is based on the estimated amount of time it takes the average student to complete all the tasks – i.e. class attendance, independent and/or group study, laboratory work, assessments etc. - related to a par-ticular academic module or course [4, 6–9, 11], subjective workload is based on the students' perception of their workload which may or may not be related to time [4, 6, 7, 10]. Kyndt, Berghmans, Dochy and Bulckens [7] further classified subjective workload into "quantitative perceived work-load" and "qualitative perceived workload" where quantitative perceived workload deals with aspects of workload and qualitative per-ceived workload had to do with the "feelings of stress, pressure or frustration" asso-ciated with high workload. Objective workload is based on course designers' estimate of how long it should take the average student do all that is required to achieve success in a module or course. It is purely based on estimates and calculations made by people other than the students themselves. A major weakness of objective workload is that it assumes the students' ability and the expected use of their time rather than the indi-vidual student's perception of their workload [6, 12]. Thus, the subjective approach is suggested. Although the subjective accounts of workload have been criticized as being a difficult means of assessing student workload [10], it is preferred as it considers the contextual factors impacting student workload.

1.2   Time Pressure and Workload

Although it has been argued that time is not the only issue that influences student workload (e.g. [4, 6]), it nevertheless plays a pivotal role. Smith [13] reported a significant correlation between time pressure and workload. Time pressure, as it relates to workload, may not solely stem from the amount of time required for studying and academic activities. It often arises because the students have other commitments that compete for their time. As Chambers [10] explains: "when students suffer interruptions to their studies as a result of illness, family difficulties or whatever, their anxiety is often expressed as a feeling of overburden."

Other factors could include part-time work, extracurricular activities and socializing with friends [6]. While it may seem quite intuitive for students to desist from or reduce participation in these activities, it may not be a realistic solution as they have their own advantages and improve their student experience. For instance, part-time work could help indigent students cater for their financial needs. Likewise, there is evidence revealing that extracurricular activities could help form strong social bonds which are important to students' overall well-being [6]. Furthermore, socializing with friends from the same class in addition to yielding social support has also been found to help students with difficult problems revealed during individual study [6]. Conversely, having limited time to spend with friends and family as a result of heavy workload has led to stress for some students [14, 15]. Therefore, even though time is not the only component of workload, it is very pivotal to its perception. Particularly, time pressure, having to do with students "having too many things to do at once", which may or may not be related to their university work is very central to their perception of their workload. Thus, making time pressure a very important issue to study.

1.3   Workload, Time Pressure and Well-Being

Firstly, it is pertinent to attempt an explanation of the well-being concept. Ryan and Deci [16] define well-being as "optimal psychological function and experience". The implication of this is that well-being is not just about living, feeling or being well, it also has to do with living to one's full potential. Well-being is divided into positive (e.g. happiness, positive affect etc.) or negative aspects (e.g. stress, depression and anxiety). There is a tendency for the well-being discourse or research about it to focus on either aspect – and thus infer well-being from the presence or absence of that aspect. This viewpoint assumes that positive and negative well-being are diametrically opposed. However, previous research (e.g. [16–18]) has revealed that this is not necessarily the case. Therefore, it has been suggested that well-being should be looked at holistically, integrating both the positive and negative aspects. Thus, helping to paint a, hopefully, more realistic picture of an individual's well-being. To measure well-being, this way, a model that integrates both aspects of well-being is required. One such model is the Demand-Resources Individual Effects (DRIVE) model [19]. It places emphasis on the individual [19] and has been used in occupational as well as educational settings (e.g. [20–22]). In addition to its being able to measure well-being holistically, the DRIVE model was not developed as "a predictive model but rather a theoretical framework into which any relevant variables can be introduced" [19]. Thus, making it flexible and adaptable to whatever well-being or related variables are to be investigated [23]. As a result of this flexibility, the DRIVE model has been used to investigate the relationship between fatigue and heavy workload in occupational samples where high workload was found to be linked with fatigue and impaired performance [3, 24]. However, to the best of our knowledge, the model has only been used by one study [13] to investigate the relationship between time pressure and/or workload and well-being in student settings. This section explores the nexus, from literature, between time pressure and/or workload on the one hand and well-being on the other. Most of the reviewed literature, except Smith (submitted), considered the relationship between either time pressure and well-being or student workload and well-being. Additionally, there was a tendency for the literature to use the typical approach of considering only one aspect of well-being. As mentioned earlier, Smith [13] found a significant correlation between time pressure and workload. This finding shows that, although both concepts are strongly related, they are not the same. It seems to align with the previous assertion that time pressure is a crucial aspect of student workload. This difference was further proven by the fact that although workload significantly predicted work efficiency, course stress, and both negative and positive well-being outcomes, time pressure only significantly predicted only course stress and negative well-being. High workload was consistently related to increased negative well-being

[13–15, 25]. It is also interesting to note that some studies [13, 14] actually showed positive relationships between high workload and positive well-being outcomes. This finding could be surprising as high workload might be thought to be negative as it places the students under high pressure and could lead to "difficulty, stress and anxiety and the intention to quit" [4]. This seems to indicate that high workload, in itself, may not be bad or be the problem but it could have some negative implications. One key implication may be the pressure it places on the individual's time resources. For instance, Bergin and Pakenham [14] found that although the respondents in this study were able to cope with their workload, the high workload meant that they were isolated socially which ultimately led to reduced life satisfaction. Furthermore, Smith [13] found time pressure to significantly predict negative well-being outcomes and course stress. Likewise, time pressure as a component of job demands was found to significantly predict anxiety and depression [25] This somewhat agrees with the findings of Kyndt, Berghmans, Dochy and Bulckens [7] who reported that some students' perception of their workload was shaped by time pressure. Although the literature linking student workload and/or time pressure on the one hand and well-being on the other seem to be reporting similar findings – a tendency for an increase in negative well-being and decrease in positive well-being – there seem to be some weaknesses in how the outcomes where characterized. One key weakness was that much of the literature (e.g. [14, 25]) limited the negative outcomes to anxiety and depression which were considered singly as opposed to Smith's approach which was a single score for the negative outcomes (comprising the sum for scores from stress, depression and anxiety). A similar trend was observed for positive well-being outcomes. For instance, Bergin and Pakenham [14] only used life satisfaction in comparison to Smith's aggregation of life satisfaction, positive affect into an overall positive outcome score. One study [25], used a student-adapted version of the Decision-Control Support (DCS) framework [26, 27]. The DCS works on the assumption that situations of high demand, low control and low social support will lead to negative consequences in terms of psychological strain. Although this study [25] findings agree with those from other studies which applied other theoretical frameworks, the DCS model has been criticized for not taking the actual individual into account [19, 28]. It works on the assumption that once the environmental characteristics are present the result is inevitable, neglecting the role that individual coping mechanisms play. This is as opposed to frameworks like the DRIVE model which attempt to account for the roles that individual differences play in the well-being process. The current study will be similar to Smith [13] in many respects. However, one key difference is that the role of nationality will be investigated in addition to those of the predictors.

1.4 Nationality, Workload, Time Pressure and Well-Being

To add another layer to the investigation of the role of workload and/or time pressure in the prediction of well-being outcomes, the role of nationality in this process will be examined. It appears none of the studies placed a very high emphasis on the role of nationality in the process. As could be expected, international students may experience study difficulties due to differences in the educational systems in their home countries and their study country [29]. However, high workload seems to be common to both international and home students [30, 31]. Although, these findings seem to contradict those of Cotton, Dollard and de Jonge [27] who found country of birth to be significantly correlated to "study load". In this study [27], study load appeared to refer to whether the student was in full-time or part-time study. Again, Sheldon and Krieger [32] controlling for ethnicity among other demographic factors revealed a general tendency for reduced positive well-being and increased negative well-being in all participants irrespective of their ethnic background and demographic characteristics. Putting all these together, it thus could be inferred that cultural differences, in terms of nationality or ethnicity do not intervene in the prediction of well-being outcomes by workload or time pressure.

1.5   Study Aim

The key aim of this study is to investigate the roles, if any, of cultural differences (nationality/ethnicity) singly, and in tandem with time pressure in the prediction of well-being outcomes as well as academic stress and work efficiency.

2   Methodology

2.1   Sample Description

The sample comprised 360 university undergraduates aged 17–42 (mean age = 20.94) from the three distinct ethnic/cultural backgrounds: White British (schooling in the United Kingdom, UK), 144; Ethnic Minorities (schooling in the UK), 103 and Nigerian (schooling in Nigeria), 113. Females made up 59.4% of the sample.

2.2   Instrument

The Qualtrics platform. Each participant gave informed consent before going on to complete the questionnaire and each phase of the data collection was approved by the Cardiff University School of Psychology Ethics Committee. The Questionnaire was delivered in the English Language. The data collected using the Student Well-being Process Questionnaire (Student WPQ) [33]. The instrument was developed to measure well-being predictors and outcomes peculiar the university undergraduate population outcomes [33] and was based on the DRIVE model [19]. The questionnaire consisted of questions encompassing various facets of well-being predictors and outcomes specific to university undergraduates. The single items were statements with explanatory statements in parentheses [34]. Some of the questions were drawn from the single-item adaptations following factor analyses of the Inventory of College Students' Recent Life Experiences (ICSRLE) [35] by Bdenhorn, Miyazki, Ng and Zalaquett [36]. An example of a single item adapted was *"Time Pressure (e.g. too many things to do at once, interruptions of your school work, a lot of responsibilities)"*. Respondents were to give ratings of 1–10 (1 = Low, 10 = High) of how the statements represented their personal experiences. In like manner, three single-items on social support were drawn from the Interpersonal Support Evaluation List (ISEL) factors short version [37].

A sample question is presented thus: *"There is a person in my life who would provide tangible support for me when I need it (e.g. money for tuition or books, use of their car, furniture for new apartment)"*. The scoring was as described for the ICSRLE, with respondents giving rating for 1–10 (1 = Low, 10 = High). Other questions covered well-being predictors covered positive personality, negative coping and conscientiousness. Most questions followed the visual scale (i.e. 1–10) which are preferred to the Likert-scale as the former has been found to be more suitable for single-item questions representing whole constructs [34]. The scores for positive personality were derived from the sum of individual scores on single-item questions for optimism, self-esteem and self-efficacy. Similarly, negative coping was an aggregate of scores on single items on avoidance, self-blame and wishful thinking which have been identified as negative coping styles. These predictors i.e. Student stressors (ICSRLE factors), social support (ISEL factors), conscientiousness, positive personality and negative coping are jointly referred to as the 'established predictors' of student well-being. They are so called because they typically predict negative and positive well-being outcomes when the student WPQ is used (e.g. [21, 22, 33]). The full questionnaire is shown in the Appendix.

2.3 Statistical Analyses

The goal of the research described here was to investigate the role played by time pressure and nationality in predicting positive well-being outcomes, negative well-being outcomes, course stress and work efficiency. While negative and positive outcomes were composites derived from summing up scores from individual outcomes, course stress and work efficiency scores were derived from the respective single item for either construct. The first step in the analyses was to split the predictor variable scores into "high" and "low" percentile groups at the median. Nationality was a nominal variable split into White British, Ethnic Minority and Nigerian based on the nationality/ethnicity of the study participants. Analysis of Variance (ANOVA) tests were then conducted to ascertain the predictive influence of time pressure and nationality on the outcomes. Time pressure and nationality were inputted as the predictors (fixed factors) while controlling for negative coping, positive personality, conscientiousness, student social support (ISEL), (academic) developmental challenges, social annoyances, academic dissatisfaction, romantic problems, social mistreatment and friendship problems (as covariates). The covariates were also split into high and low groups at the median. Also, worth mentioning is that (academic) developmental challenges, academic dis-satisfaction, romantic problems, social mistreatment were factors from the ICSRLE from which time pressure emerged. The main effects of each of the predictors (time pressure and nationality) and co-variates in the prediction of the outcomes were investigated. Likewise, the two-way interactions between nationality and time pressure, as well as with each of the co-variates. For instance, the predictive effect of academic dissatisfaction in predicting the outcomes was investigated singly, as well as the predictive effects of academic dissatisfaction in tandem with nationality. These analyses were done with SPSS 25.

3 Results

3.1 Negative Outcomes

ANOVAs were carried out as described in the previous section for the prediction of negative outcomes. The negative outcome scores comprised scores from individual single-items on depression, anxiety and life stress. Some of the established effects were observed: negative coping and positive personality both significantly predicted negative outcomes (p < .001). Table 1 presents all the results of the significant effects and those of nationality and time pressure.

Table 1. Prediction of Negative Outcomes (Significant Predictions only and predictions by nationality and time pressure)

Predictor		Sum of squares	df	Mean square	F	Sig.
Nationality	Between groups	626.54	2	313.27	4.67	**.010**
	Within groups	21530.32	321	67.07		
	Total	22156.86	323	380.34		
Time pressure	Between groups	186.13	1	186.13	2.76	.097
	Within groups	21530.32	321	67.07		
	Total	21716.45	322	253.2		
Negative coping	Between groups	3197.26	1	3197.26	47.67	**.000**
	Within groups	21530.32	321	67.07		
	Total	24727.58	322	3264.33		
Positive personality	Between groups	3579.44	1	3579.44	53.37	**.000**
	Within groups	21530.32	321	67.07		
	Total	25190.76	322	3646.51		
Social mistreatment	Between groups	286.42	1	286.42	4.27	**.040**
	Within groups	21530.32	321	67.07		
	Total	21816.74	322	353.49		
Nationality*Time pressure	Between groups	479.22	2	239.61	3.57	**.029**
	Within groups	21530.32	321	67.07		
	Total	22009.54	323	306.68		
Nationality*Positive personality	Between groups	658.04	2	329.02	4.91	**.008**
	Within groups	21530.32	321	67.07		
	Total	22188.36	323	396.09		
Nationality*Academic dissatisfaction	Between groups	417.87	2	208.94	3.12	**.046**
	Within groups	21530.32	321	67.07		
	Total	21948.19	323	276.01		

Nationality was found to significantly predict negative outcomes ($p < .05$). Figure 1 shows the mean values of negative outcomes for the three groups.

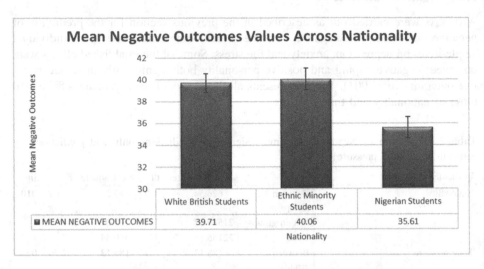

Fig. 1. Mean values of negative outcomes across nationality

Although time pressure, on its own, did not significantly predict negative outcomes, it did so when combined with nationality. Table 2 presents the means of negative outcomes as predicted by the combination of Nationality and time pressure.

Table 2. Mean of Negative Outcomes predicted by Nationality*Time pressure

	Mean (SD)	Mean (SD)
Nationality	Low time pressure	High time pressure
White British students	37.58 (1.18)	41.84 (1.26)
Ethnic minority students	41.08 (1.22)	39.04 (1.42)
Nigerian students	34.13 (1.16)	37.09 (1.50)

3.2   Positive Outcomes

Positive outcomes score was derived by summing up the responses to single-item questions on happiness, life satisfaction and positive affect. The findings show that neither nationality nor time pressure significantly predicted positive outcomes, although nationality combined with social annoyance yielded significant predictions. Furthermore, positive personality, student social support and conscientiousness were all significant predictors of positive outcomes. These findings are presented in Table 3.

Table 3. Prediction of Positive Outcomes (Significant Predictions only and predictions by nationality and time pressure)

Predictor		Sum of squares	df	Mean square	F	Sig.
Nationality	Between groups	11.27	2	5.64	.418	.659
	Within groups	4328.80	321	13.49		
	Total	4340.07	323	19.13		
Time pressure	Between groups	.09	1	.09	.007	.935
	Within groups	4328.80	321	13.49		
	Total	4328.89	322	13.58		
Positive personality	Between groups	293.85	1	293.85	21.79	**.000**
	Within groups	4328.80	321	13.49		
	Total	4622.35	322	307.34		
Conscientiousness	Between groups	90.80	1	90.80	6.73	**.010**
	Within groups	4328.80	321	13.49		
	Total	4419.60	322	104.29		
Student social support	Between groups	406.40	1	406.40	30.14	**.000**
	Within groups	4328.80	321	13.49		
	Total	4735.20	322	419.89		
Nationality*Social annoyance	Between groups	189.16	2	94.58	7.01	.001
	Within groups	4328.80	321	13.49		
	Total	4517.96	323	108.07		
Nationality*Time pressure	Between groups	25.10	2	12.55	.931	.395
	Within groups	4328.80	321	13.49		
	Total	4353.9	323	26.04		

3.3   Course Stress

Course stress was measured by a single item which asked the respondents to give a rating of 1–10 (1 = Low, 10 = High) of how stressful they found their course to be. Time pressure significantly predicted course stress ($p < .001$), as did the combination of nationality and positive personality ($p < .05$). There were no other significant predictions observed. The findings are presented in Table 4. Figure 2 also presents the means of high and low time in the prediction of course stress.

Table 4. Prediction of Course Stress (Significant Predictions only and predictions by nationality and time pressure)

Predictor		Sum of squares	df	Mean square	F	Sig.
Nationality	Between groups	8.14	2	4.07	1.274	.281
	Within groups	1025.87	321	3.20		
	Total	1034.01	323	7.27		
Time pressure	Between groups	52.13	1	52.13	16.31	**.000**
	Within groups	1025.87	321	3.20		
	Total	1078	322	55.33		
Nationality*Positive personality	Between groups	22.46	2	11.23	3.51	**.031**
	Within groups	1025.87	321	3.20		
	Total	1048.33	323	14.43		
Nationality*Time pressure	Between groups	18.17	2	9.09	2.844	.060
	Within groups	1025.87	321	3.20		
	Total	1044.05	323	12.29		

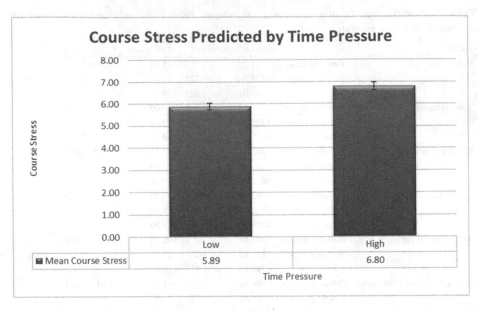

Fig. 2. Course stress predicted by time pressure

3.4 Work Efficiency

Work efficiency was a single-item subjective measure of how efficiently the student carried out their university work. Developmental challenges, conscientiousness, student social support all significantly predicted work efficiency. Also, when nationality combined with time pressure, they significantly predicted work efficiency. Table 5 presents the results in detail. Table 6 presents the means for work efficiency for high and low time pressure according to nationality.

Table 5. Prediction of Work Efficiency (Significant Predictions only and predictions by nationality and time pressure)

Predictor		Sum of squares	df	Mean square	F	Sig.
Nationality	Between groups	8.67	2	4.33	1.133	.323
	Within groups	1227.97	321	3.83		
	Total	1236.64	323	8.16		
Time pressure	Between groups	.93	1	.93	.244	.622
	Within groups	1227.97	321	3.83		
	Total	1228.9	322	4.76		
Developmental problems	Between groups	17.43	1	17.43	4.56	**.034**
	Within groups	1227.97	321	3.83		
	Total	1245.40	322	21.26		
Student social support	Between groups	21.67	1	21.67	5.66	**.018**
	Within groups	1227.97	321	3.83		
	Total	1249.64	322	25.5		
Nationality*Time pressure	Between groups	31.29	2	15.65	4.09	**.018**
	Within groups	1227.97	321	3.83		
	Total	1259.26	323	19.48		

Table 6. Means of work efficiency predicted by Nationality*Time pressure

	Mean (SD)	Mean (SD)
Nationality	Low time pressure	High time pressure
White British	5.23 (.28)	5.49 (.30)
Ethnic minority	6.49 (.29)	7.00 (.34)
Nigerian	7.69 (.28)	6.56 (.36)

4 Discussion

This study investigated the role of cultural differences (nationality, ethnicity) and time pressure in the prediction of well-being outcomes as well as work efficiency and course stress while controlling for the established factors (negative coping, positive personality, social support (ISEL factors), conscientiousness) and all the other ICSRLE student stressors except time pressure. The first thing worth noting was that some of the established effects for the prediction of negative and positive well-being outcomes were retained. This agrees with previous research with the student WPQ/DRIVE [21, 33]. However, nationality on its own was found to significantly predict negative outcomes, with the Nigerian students showing much lower scores than the White British and Ethnic Minority Students (Fig. 1). This contradicts the findings of Omosehin and Smith [38] which found nationality, on its own, to not predict either positive or negative outcomes in workers. This could be indicating that the well-being process in students differs quite markedly from that of workers. Another plausible explanation is that, as the mean negative outcomes score for Nigerian students was much lower than those of the others (Fig. 1), it could be that nationality per se does not necessarily predict negative outcomes but, rather, the extent to which it is experienced. In that wise, it could be argued that the Nigerian students experienced negative outcomes to a much lesser degree. Yet another possibility is that as the British sample (White British and Ethnic Minority samples) experienced higher negative outcomes than the Nigerian sample, it could be that some geographical or environmental factors affected their perception of experiencing negative well-being. Nigerian students probably perceived some issues as being normal aspects of everyday living or student life and did not perceive them as negative well-being issues. This may also have to do with resilience in terms of seeing an issue as a challenge to be overcome or a problem to be solved rather than a negative situation. Resilience is an individual's ability to "thrive in the face of adversity" [39] and this seems to aptly describe the Nigerian students' disposition to perceived negative circumstances. Future studies should include larger samples and include variables like resilience in order to further investigate this. Qualitative studies should also be encouraged in order to tease out possible contextual factors that could account for this disparity in perceptions. Time pressure is not the same thing as workload. Nevertheless, it plays a pivotal role as time is crucial to the perception of workload. This was further buttressed by Smith [13] who found a significant correlation between time pressure and workload. Therefore, it follows that perceptions of time pressure will greatly influence workload. The current study found time pressure to significantly predict course stress, which replicates Smith's [14] findings. This finding seems logical as students who feel they do not have enough time to give to their academic activities are likely to view their course as stressful. The feelings that the course is stressful could increase the likelihood of plagiarism. Devlin and Gray [8] report that time pressure leads to plagiarism. This may be a result of their inability to manage their time effectively to juggle academic and non-academic aspects of their lives. This is somewhat in line with Chambers [10] who opines that "when students suffer interruptions to their studies as a result of illness, family difficulties or whatever, their anxiety is often expressed as a feeling of overburden." Furthermore, time pressure, in tandem with nationality predicted negative outcomes and work efficiency. These findings can, however, can be criticized as being chance or spurious effects, owing to the small effect sizes, partial eta squared (not presented). Furthermore, if corrections are made for multiple comparisons, the p values are likely to be no longer significant. These cast doubts on these findings and calls for further research with much larger samples. Finally, this study showed that nationality did not predict positive outcomes and this finding is in concord with findings from a cross-cultural occupational study [38].

4.1 Limitations

It has been stated in this study that time pressure is not synonymous with student workload. Although, previous research has established significant links between the two concepts and literature has shown that time is very pivotal to how workload is perceived, they are quite different. The implication of this difference, especially for the current study, is that time pressure can only present an aspect of the effects of workload on well-being. Particularly in this situation where the roles of nationality were also investigated, an actual measure(s) of workload would have given a clearer picture of the effects of nationality and workload on well-being. Furthermore, this study only relied on subjective outcomes. While subjective perceptions of well-being are very important as they are relative from one individual to another, it would have been helpful to include measure like students' academic performance (in terms of Grade Point Average GPA). To this end, it is suggested that future studies should incorporate actual measures of workload and objective outcomes (like academic attainment as measured by GPA) addition to nationality, time pressure and other variables (predictors and outcomes) investigated in the current study. Another limitation is the cross-sectional nature of the study which makes it impossible to established causality. It is thus suggested that future studies should be longitudinal in nature while incorporating all the aforementioned variables.

5 Conclusion

Time pressure is very central to student workload. This study has revealed that time pressure, on its own, predicted course stress. Time pressure in tandem with nationality predicted work efficiency and negative well-being outcomes. Time pressure directly or indirectly predicted three of the four outcomes measured. As time pressure plays a huge role in student workload, it is a critical issue that needs to be managed to optimize student well-being. Furthermore, nationality predicted negative well-being out-comes, and in tandem with time pressure and other student well-being predictors predicted all the outcomes. This indicates that cultural differences do play significant roles in the prediction of well-being especially when combined with predictors - emphasizing the contextual factors that influence student well-being.

Appendix

Student Well-being Questionnaire

The following questions contain a number of single-item measures of aspects of your life as a student and feelings about yourself. Many of these questions will contain examples of what thoughts/behaviours the question is referring to which are important for understanding the focus of the question, but should be regarded as guidance rather than strict criteria. Please try to be as accurate as possible, but avoid thinking too much about your answers, your first instinct is usually the best.

1. Overall, I feel that I have low self-esteem (For example: At times, I feel that I am no good at all, at times I feel useless, I am inclined to feel that I am a failure)

   Disagree strongly 1 2 3 4 5 6 7 8 9 10 Agree strongly

2. On a scale of one to ten, how depressed would you say you are in general? (e.g. feeling 'down', no longer looking forward to things or enjoying things that you used to)

   Not at all depressed 1 2 3 4 5 6 7 8 9 10 Extremely depressed

3. I have been feeling good about my relationships with others (for example: Getting along well with friends/colleagues, feeling loved by those close to me)

   Disagree strongly 1 2 3 4 5 6 7 8 9 10 Agree strongly

4. I don't really get on well with people (For example: I tend to get jealous of others, I tend to get touchy, I often get moody)

   Disagree strongly 1 2 3 4 5 6 7 8 9 10 Agree strongly

5. Thinking about myself and how I normally feel, in general, I mostly experience positive feelings (For example: I feel alert, inspired, determined, attentive)

   Disagree strongly 1 2 3 4 5 6 7 8 9 10 Agree strongly

6. In general, I feel optimistic about the future (For example: I usually expect the best, I expect more good things to happen to me than bad, It's easy for me to relax)

   Disagree strongly 1 2 3 4 5 6 7 8 9 10 Agree strongly

7. I am confident in my ability to solve problems that I might face in life (For example: I can usually handle whatever comes my way, If I try hard enough I can overcome difficult problems, I can stick to my aims and accomplish my goals)

Disagree strongly 1 2 3 4 5 6 7 8 9 10 Agree strongly

8. I feel that I am laid-back about things (For example: I do just enough to get by, I tend to not complete what I've started, I find it difficult to get down to work)

Disagree strongly 1 2 3 4 5 6 7 8 9 10 Agree strongly

9. I am not interested in new ideas (For example: I tend to avoid philosophical discussions, I don't like to be creative, I don't try to come up with new perspectives on things)

Disagree strongly 1 2 3 4 5 6 7 8 9 10 Agree strongly

10. Overall, I feel that I have positive self-esteem (For example: On the whole I am satisfied with myself, I am able to do things as well as most other people, I feel that I am a person of worth)

Disagree strongly 1 2 3 4 5 6 7 8 9 10 Agree strongly

11. I feel that I have the social support I need (For example: There is someone who will listen to me when I need to talk, there is someone who will give me good advice, there is someone who shows me love and affection)

Disagree strongly 1 2 3 4 5 6 7 8 9 10 Agree strongly

12. Thinking about myself and how I normally feel, in general, I mostly experience negative feelings (For example: I feel upset, hostile, ashamed, nervous)

Disagree strongly 1 2 3 4 5 6 7 8 9 10 Agree strongly

13. I feel that I have a disagreeable nature (For example: I can be rude, harsh, unsympathetic)

Disagree strongly 1 2 3 4 5 6 7 8 9 10 Agree strongly

Negative Coping Style
Blame Self

14. When I find myself in stressful situations, I blame myself (e.g. I criticize or lecture myself, I realise I brought the problem on myself).

Disagree strongly 1 2 3 4 5 6 7 8 9 10 Agree strongly

Wishful Thinking

15. When I find myself in stressful situations, I wish for things to improve (e.g. I hope a miracle will happen, I wish I could change things about myself or circumstances, I daydream about a better situation).

Disagree strongly 1 2 3 4 5 6 7 8 9 10 Agree strongly

Avoidance

16. When I find myself in stressful situations, I try to avoid the problem (e.g. I keep things to myself, I go on as if nothing has happened, I try to make myself feel better by eating/drinking/smoking).

    Disagree strongly 1 2 3 4 5 6 7 8 9 10 Agree strongly

Personality

17. I prefer to keep to myself (For example: I don't talk much to other people, I feel withdrawn, I prefer not to draw attention to myself)

    Disagree strongly 1 2 3 4 5 6 7 8 9 10 Agree strongly

18. I feel that I have an agreeable nature (For example: I feel sympathy toward people in need, I like being kind to people, I'm co-operative)

    Disagree strongly 1 2 3 4 5 6 7 8 9 10 Agree strongly

19. In general, I feel pessimistic about the future (For example: If something can go wrong for me it will, I hardly ever expect things to go my way, I rarely count on good things happening to me)

    Disagree strongly 1 2 3 4 5 6 7 8 9 10 Agree strongly

20. I feel that I am a conscientious person (For example: I am always prepared, I make plans and stick to them, I pay attention to details)

    Disagree strongly 1 2 3 4 5 6 7 8 9 10 Agree strongly

21. I feel that I can get on well with others (For example: I'm usually relaxed around others, I tend not to get jealous, I accept people as they are)

    Disagree strongly 1 2 3 4 5 6 7 8 9 10 Agree strongly

22. I feel that I am open to new ideas (For example: I enjoy philosophical discussion, I like to be imaginative, I like to be creative)

    Disagree strongly 1 2 3 4 5 6 7 8 9 10 Agree strongly

23. Overall, I feel that I am satisfied with my life (For example: In most ways my life is close to my ideal, so far I have gotten the important things I want in life)

    Disagree strongly 1 2 3 4 5 6 7 8 9 10 Agree strongly

24. On a scale of one to ten, how happy would you say you are in general?

    Extremely unhappy 1 2 3 4 5 6 7 8 9 10 Extremely happy

25. On a scale of one to ten, how anxious would you say you are in general? (e.g. feeling tense or 'wound up', unable to relax, feelings of worry or panic)

    Not at all anxious 1 2 3 4 5 6 7 8 9 10 Extremely anxious

26. Overall, how stressful is your life?

Not at all stressful 1 2 3 4 5 6 7 8 9 10 Very Stressful

Please consider the following elements of student life and indicate overall to what extent they have been a part of your life over the past 6 months. Remember to use the examples as guidance rather than trying to consider each of them specifically:

27. Challenges to your development (e.g. important decisions about your education and future career, dissatisfaction with your written or mathematical ability, struggling to meet your own or others' academic standards).

Not at all part of my life 1 2 3 4 5 6 7 8 9 10 Very much part of my life

28. Time pressures (e.g. too many things to do at once, interruptions of your school work, a lot of responsibilities).

Not at all part of my life 1 2 3 4 5 6 7 8 9 10 Very much part of my life

29. Academic Dissatisfaction (e.g. disliking your studies, finding courses uninteresting, dissatisfaction with school).

Not at all part of my life 1 2 3 4 5 6 7 8 9 10 Very much part of my life

30. Romantic Problems (e.g. decisions about intimate relationships, conflicts with boyfriends'/girlfriends' family, conflicts with boyfriend/girlfriend).

Not at all part of my life 1 2 3 4 5 6 7 8 9 10 Very much part of my life

31. Societal Annoyances (e.g. getting ripped off or cheated in the purchase of services, social conflicts over smoking, disliking fellow students).

Not at all part of my life 1 2 3 4 5 6 7 8 9 10 Very much part of my life

32. Social Mistreatment (e.g. social rejection, loneliness, being taken advantage of).

Not at all part of my life 1 2 3 4 5 6 7 8 9 10 Very much part of my life

33. Friendship problems (e.g. conflicts with friends, being let down or disappointed by friends, having your trust betrayed by friends).

Not at all part of my life 1 2 3 4 5 6 7 8 9 10 Very much part of my life
Please state how much you agree or disagree with the following statements:

34. There is a person or people in my life who would provide tangible support for me when I need it (for example: money for tuition or books, use of their car, furniture for a new apartment).

Strongly Disagree 1 2 3 4 5 6 7 8 9 10 Strongly Agree

35. There is a person or people in my life who would provide me with a sense of belonging (for example: I could find someone to go to a movie with me, I often get invited to do things with other people, I regularly hang out with friends).

Strongly Disagree 1 2 3 4 5 6 7 8 9 10 Strongly Agree

36. There is a person or people in my life with whom I would feel perfectly comfortable discussing any problems I might have (for example: difficulties with my social life, getting along with my parents, sexual problems).

Strongly Disagree 1 2 3 4 5 6 7 8 9 10 Strongly Agree

References

1. Wickens, C.D., Hollands, J.G., Banbury, S., Parauraman, R.: Engineering Psychology and Human Performance, 4th edn. Routledge, Abingdon (2016)
2. Byrne, A.: The effect of education and training on mental workload in medical education. In: Longo, L., Leva, M.Chiara (eds.) H-WORKLOAD 2018. CCIS, vol. 1012, pp. 258–266. Springer, Cham (2019). https://doi.org/10.1007/978-3-030-14273-5_15
3. Fan, J., Smith, A.P.: The impact of workload and fatigue on performance. In: Longo, L., Leva, M.C. (eds.) H-WORKLOAD 2017. CCIS, vol. 726, pp. 90–105. Springer, Cham (2017). https://doi.org/10.1007/978-3-319-61061-0_6
4. Bowyer, K.: A model of student workload. J. High. Educ. Policy Manag. **34**(3), 239–258 (2012). https://doi.org/10.1080/1360080x.2012.678729
5. Kingsland, A.J.: Time expenditure, workload, and student satisfaction in problem-based learning. New Dir. Teach. Learn. **1996**(68), 73–81 (1996). https://doi.org/10.1002/tl.37219966811
6. Kember, D.: Interpreting student workload and the factors which shape students' perceptions of their workload. Stud. High. Educ. **29**(2), 165–184 (2004). https://doi.org/10.1080/0307507042000190778
7. Kyndt, E., Berghmans, I., Dochy, F., Bulckens, L.: 'Time is Not Enough.' Workload in Higher Education: a student perspective. High. Educ. Res. Dev. **33**(4), 684–698 (2013). https://doi.org/10.1080/07294360.2013.863839
8. Devlin, M., Gray, K.: In their own words: a qualitative study of the reasons Australian university students plagiarize. High. Educ. Res. Dev. **26**(2), 181–198 (2007). https://doi.org/10.1080/07294360701310805
9. Baeten, M., Kyndt, E., Struyven, K., Dochy, F.: Using student-centred learning environments to stimulate deep approaches to learning: factors encouraging or discouraging their effectiveness. Educ. Res. Rev. **5**(3), 243–260 (2010). https://doi.org/10.1016/j.edurev.2010.06.001
10. Chambers, E.: Work-load and the quality of student learning. Stud. High. Educ. **17**(2), 141–153 (1992). https://doi.org/10.1080/03075079212331382627
11. Souto-Iglesias, A., Baeza_Romero, M.: Correction to: a probabilistic approach to student workload: empirical distributions and ECTS. High. Educ. **76**(6), 1027–1027 (2018). https://doi.org/10.1007/s10734-018-0270-1
12. Ramsden, P.: Improving teaching and learning in higher education: the case for a relational perspective. Stud. High. Educ. **12**(3), 275–286 (1987). https://doi.org/10.1080/03075078712331378062
13. Smith, A.P.: Student Workload, Wellbeing and Academic Attainment Submitted
14. Bergin, A., Pakenham, K.: Law student stress: relationships between academic demands, social isolation, career pressure, study/life imbalance and adjustment outcomes in law students. Psychiatry Psychol. Law **22**(3), 388–406 (2014). https://doi.org/10.1080/13218719.2014.960026

15. Pritchard, M., McIntosh, D.: What predicts adjustment among law students? A longitudinal panel study. J. Soc. Psychol. **143**(6), 727–745 (2003). https://doi.org/10.1080/00224540309 600427
16. Ryan, R.M., Deci, E.L.: On happiness and human potentials: a review of research on hedonic and eudaimonic well-being. Annu. Rev. Psychol. **52**, 141–166 (2001). https://doi.org/10. 1146/annurev.psych.52.1.141
17. Caccioppo, J.T., Bernsto, G.B.: The affect system: architecture and operating characteristics. Curr. Dir. Psychol. Sci. **8**, 133–137 (1999). https://doi.org/10.1111/1467-8721.00031
18. Smith, A.P., Wadsworth, E.A.: A holistic approach to stress and wellbeing. Part 5: what is a good job? Occup. Health (At work) **8**, 25–27 (2011)
19. Mark, G., Smith, A.P.: Stress models: a review and suggested new direction. In: Houdmont, J., Leka, S. (eds.) Occupational Health Psychology, vol. 3, pp. 111–144. Nottingham University Press, Nottingham (2008)
20. Zurlo, M.C., Vallone, F., Smith, A.P.: Effects of individual differences and job characteristics on the psychological health of Italian nurses. Europe's J. Psychol. **14**, 159–175 (2018). https://doi.org/10.5964/ejop.v14i1.1478
21. Smith, A.P.: Cognitive fatigue and the wellbeing and academic attainment of university students. J. Educ. Soc. Behav. Sci. **24**(2), 1–12 (2018). https://doi.org/10.9734/jesbs/2018/ 39529
22. Williams, G.M., Smith, A.P.: A longitudinal study of the well-being of students using the student wellbeing process questionnaire (Student WPQ). J. Educ. Soc. Behav. Sci. **24**(4), 1–6 (2018). https://doi.org/10.9734/jesbs/2018/40105
23. Omosehin, O., Smith, A.P.: Adding new variables to the well-being process questionnaire (WPQ) – further studies of workers and students. J. Educ. Soc. Behav. Sci. **28**(3), 1–19 (2018). https://doi.org/10.9734/jesbs/2018/45535
24. Smith, A.P., Smith, H.N.: Workload, fatigue and performance in the rail industry. In: Longo, L., Leva, M.C. (eds.) H-WORKLOAD 2017. CCIS, vol. 726, pp. 251–263. Springer, Cham (2017). https://doi.org/10.1007/978-3-319-61061-0_17
25. Chambel, M.J., Curral, L.: Stress in academic life: work characteristics as predictors of student well-being and performance. Appl. Psychol. **54**(1), 135–147 (2005). https://doi.org/ 10.1111/j.1464-0597.2005.00200.x
26. Karasek, R., Theorell, T.: Healthy Work: Stress, Productivity and the Reconstruction of Working Life. Basic Books, New York (1990)
27. Cotton, S.J., Dollard, M.F., de Jonge, J.: Stress and job design: satisfaction, well-being and performance in university students. Int. J. Stress. Manag. **9**(3), 147–162 (2002)
28. Perrewé, P.L., Zellars, K.L.: An examination of attributions and emotions in the transactional approach to the organizational stress process. J. Organ. Behav. **20**(5), 739–752 (1999). https://doi.org/10.1002/(sici)1099-1379(199909)20:5%3c739:aid-job1949%3e3.0.co;2-c
29. Ballard, B.: Academic adjustment: the other side of the export dollar. High. Educ. Res. Dev. **6**(2), 109–119 (1987). https://doi.org/10.1080/0729436870060203
30. Mullins, G., Quintrell, N., Hancock, L.: The experiences of international and local students at three australian universities. High. Educ. Res. Dev. **14**(2), 201–231 (1995). https://doi.org/ 10.1080/0729436950140205
31. Alharbi, E., Smith, A.: Review of the literature on stress and wellbeing of international students in English-speaking countries. Int. Educ. Stud. **11**(6), 22 (2018). https://doi.org/10. 5539/ies.v11n6p22
32. Sheldon, K.M., Krieger, L.S.: Does legal education have undermining effects on law students? Evaluating changes in motivation, values, and well-being. Behav. Sci. Law **22**(2), 261–286 (2004). https://doi.org/10.1002/bsl.582

33. Williams, G., Pendlebury, H., Thomas, K., Smith, A.: The student well-being process questionnaire (student WPQ). Psychology **08**(11), 1748–1761 (2017). https://doi.org/10.4236/psych.2017.811115

34. Williams, G., Thomas, K., Smith, A.: Stress and well-being of university staff: an investigation using the demands-resources-individual effects (DRIVE) model and well-being process questionnaire (WPQ). Psychology **08**(12), 1919–1940 (2017). https://doi.org/10.4236/psych.2017.812124

35. Kohn, P.M., Lafreniere, K., Gurevich, M.: The Inventory of College Students' Recent Life Experiences: a decontaminated hassles scale for a special population. J. Behav. Med. **13**(6), 619–630 (1990). https://doi.org/10.1007/bf00844738

36. Bodenhorn, N., Miyazaki, Y., Ng, K., Zalaquett, C.: Analysis of the inventory of college students' recent life experiences. Multicult. Learn. Teach. **2**(2) (2007). https://doi.org/10.2202/2161-2412.1022

37. Cohen, S., Mermelstein, R., Kamarck, T., Hoberman, H.: Measuring the Functional Components of Social Support (1986)

38. Omosehin, O., Smith, A.: Nationality, ethnicity and the well-being process in occupational samples. Open J. Soc. Sci. **07**(05), 133–142 (2019). https://doi.org/10.4236/jss.2019.75011

39. Connor, K.M., Davidson, J.R.: Development of a new resilience scale: the Connor-Davidson Resilience Scale (CD-RISC). Depress. Anxiety **18**(2), 76–82 (2003). https://doi.org/10.1002/da.10113

Ocular Indicators of Mental Workload: A Comparison of Scanpath Entropy and Fixations Clustering

Piero Maggi[✉], Orlando Ricciardi[✉], and Francesco Di Nocera[✉]

Department of Psychology, Sapienza University of Rome,
Via dei Marsi, 78 - 00185 Rome, Italy
{piero.maggi, orlando.ricciardi,
francesco.dinocera}@uniroma1.it

Abstract. Among the different eye-tracking metrics, both the entropy-based analysis of the scanpath and the spatial distribution of fixations points have been suggested as indices of mental workload. However, they have never been directly compared so far. In this study, eye movements were recorded from fourteen subjects while performing a visuo-spatial task (the "spot the differences" puzzle game). Subjective workload ratings were also collected throughout the session along with performance data. Entropy Rate and Nearest Neighbor Index show matching patterns both indicating high mental workload after two minutes of visual scanning. As predicted elsewhere, the distribution of fixations becomes more clustered under increases of the visuo-spatial demand and showing the same pattern of the Entropy rate that becomes more stereotyped. Temporal demand, instead, has been reported to produce spreading of the fixation pattern. Interestingly, both ocular indices seem to anticipate the drop in performance that is visible after the sixth minute of visual scanning.

Keywords: Mental workload · Eye-tracking · Scanpath · Fixations · Entropy

1 Introduction

The relation between mental workload and ocular activity has been reported many times and using different metrics (e.g. pupil diameter, fixations duration, saccades amplitude, blink rate) [1–4]. However, one interesting and not much-pursued way of taking into account eye movements is to analyze the overall sequence of fixations and saccades, the so-called "scanpath". The topography of the visual scanning, as well as its dynamics, was quantitatively approached in two studies by Tole and colleagues [5] and Harris and colleagues [6] who suggested to use the entropy rate of the visual scanning for discriminating between different levels of mental workload. Their results suggested that visual scanning would tend to disorder when the workload is low, while it would become more stereotyped as the demand increases. Although very appealing, entropy has been seldom used as a measure of workload since then and its properties have not been properly tested. Moreover, entropy is limited by the need to rely on predefined Areas Of Interest (AOIs) in order to compute transitions between them: in many operational settings visual scanning happens outside specific AOIs, or the AOIs change

© Springer Nature Switzerland AG 2019
L. Longo and M. C. Leva (Eds.): H-WORKLOAD 2019, CCIS 1107, pp. 205–212, 2019.
https://doi.org/10.1007/978-3-030-32423-0_13

dynamically. For overcoming this limit, Di Nocera, Camilli and Terenzi [7] introduced an alternative approach based on the analysis of the fixations distribution and using spatial statistics for computing the Nearest Neighbor Index (NNI), which was found to be sensitive to changes in mental workload. These results have been confirmed by other studies from independent research laboratories. Earlier studies on the functional significance of this index and its sensitivity to different task demands suggested that fixations appear to be more scattered over the entire visual field when an increase in workload is due to time pressure (temporal demand), while higher concentration of fixations in specific portions of the area seems to depend on the visuospatial demand [8]. Stereotypy in visual scanning predicted by the entropy approach when workload increases may be consistent with the fixation clustering predicted by the spatial distribution approach. Entropy, indeed, is based on transitions between AOIs and measured in tasks in which the main workload component is the visuo-spatial demand. However, the two indices have never been directly compared so far. The analytical details of the two approaches is described in the following sections.

1.1   Entropy Rate

Tole [5] applied the concept of entropy, intended as a measure of the disorder present in any physical system, to eye movements. When entropy is high, the individual will look at all quadrants in the scene an approximately equal number of times and will switch between all possible combinations of stimuli while maintaining a stable frequency. When the individual begins to focus on a narrower range of potential points of interest, the entropy value will decrease. That happens because the frequency of transitions to other areas decreases. In a state of low Entropy, a systematic and more regular strategy in visual exploration will begin to emerge and as a result, the transition from one area to another will become more ordered. The main advantage of using this analysis is that it is possible to "summarize" the visual strategy used by means of a single value. To estimate the amount of entropy, the first step is to define the areas of interest within the visual field. Then, compute the proportion of time the subject has observed each of these areas:

$$\text{Entropy Rate} = \sum_{i=1}^{D} [(E/E \max)/DT_i]$$

$$E = -\sum_{i=1}^{D} P_i \log_2 P_i$$

Where E is the observed average entropy, E-max is the value of the maximum entropy calculated from the total number of AOIs present in the visual scene (it represents the value of entropy when all the AOIs are observed with equal probability), Pi is the probability of occurrence of the sequence i, DTi is the average fixation duration for the i-th sequence during visual exploration and D is the number of different sequences in the scanpath. This index is expressed in bits/second.

1.2   Nearest Neighbor Index

The Nearest Neighbor Index provides information about the spatial distribution of points. Its application to eye movements takes into consideration the average distance between fixations recorded during the execution of a task and the average distance between fixations that could be expected in a random distribution. The result is expressed as a single value where 1 indicates that the empirical distribution is no different from a random distribution and values above or below to 1 indicate a dispersion or a clustering of the pattern, respectively. Provided that enough fixations are available (about 50 as a rule of thumb), the index can be computed for small epochs and therefore analyzed as time series therefore providing information about the ongoing variations in mental workload. The validity of this algorithm as a measure of mental workload was supported by a methodological study [8] showing the consistency of the index with subjective (NASA-TLX score) and psychophysiological (amplitude of the P300 component of event-related potentials) measures. To estimate the index the first step is to compute the nearest neighbor distance or d(NN):

$$d(NN) = \sum_{i=1}^{N} \left[min \frac{(dij)}{N} \right]$$

where min (dij) is the distance between each point and the nearest point and N is the number of points in the distribution.

The second element of the equation is obtained by computing the average random distance or d(ran); this value would correspond to the value of d(NN) if the distribution of points were completely random:

$$d(ran) = 0.5 \sqrt{\frac{A}{N}}$$

where A is the area of the polygon defined by the most extreme fixations and N is the number of points. Finally, the NNI value is computed by dividing the nearest neighbor distance, d(NN), by the average random distance, d(ran):

$$NNI = \frac{d(NN)}{d(ran)}$$

2   Study

This study was aimed at contrasting, in terms of their sensitivity in reflecting variations in mental workload, the entropy-based scanpath method with a spatial statistics approach to the distribution of fixations points.

2.1   Participants

Fourteen university students (9 women and 6 males, mean age = 24 years, St. dev. = 2.6) volunteered in the experiment. All participants had a normal or corrected-to-normal vision, were naïve as to the aims of the experiment, its expected outcomes, and its methodology. This research complied with the tenets of the Declaration of Helsinki and was approved by the Institutional Review Board of the Department of Psychology, Sapienza University of Rome, Italy. Informed consent was obtained from each participant.

2.2 Stimuli

Two black and white pictures (Figs. 1 and 2) were displayed side by side horizontally in full-screen mode on a 27″ display. The size of each picture was 9.8 × 5.5 in. As in the "spot the differences" notorious game, the two pictures featured thirty-five small differences, but were otherwise identical. For inducing high visual-spatial demand and to verify how that affects visual exploration, we have chosen to use a single pair of images that are rich in details and would engage subjects in a visual search long session.

2.3 Eye Movements Recording

Participants eye movements were recorded using the Pupil Labs system with binocular 120 Hz Eye Tracking Camera (Pupil Labs GmbH, Germany). Participants performed a nine-point calibration prior to recording.

2.4 Procedure

Participants sat in a dark room at approximately 2 ft. from a computer display. They were requested to find as many differences they could find between two pictures (Figs. 1 and 2) in a 24-min session. Participants were instructed to mouse-click on each difference they found. Each difference found was kept circled for all the duration of the task. During the execution of the task, participants were prompted to provide a subjective assessment of mental workload from 1 (low) to 5 (high) on a 2-min schedule (Instantaneous Self-Assessment [9]).

3   Data Analysis and Results

3.1   Performance and Self-reporting Measures

In order to match performance and subjective ratings, the entire activity was divided into 12 epochs of two minutes each. The number of differences found by each participant in each time period was considered a performance indicator. The number of the differences found was used as dependent variables in repeated measures ANOVA design using Epoch as repeated factor. Results showed a main effect of Epoch ($F_{11,143} = 16.52$, $p < .001$; Fig. 3). Instantaneous Self Assessment rating was used as dependent variable in a repeated measures ANOVA design using Epoch as repeated factor. Results showed a main effect of Epoch ($F_{11,143} = 15.50$, $p < .001$; Fig. 3). In both cases, Duncan post-hoc testing showed a steady pattern (asymptotic performance and workload) starting from minute twelve.

Fig. 2. Right panel

Fig. 1. Left panel

3.2   Analysis of Eye-Tracking Data

Nearest Neighbor Index. The NNI is a value ranging from 0 to a maximum of 2.15, where 1 corresponds to a random pattern of fixation points. Values less than 1 indicate a clustered pattern, while values greater than 1 indicate an orderly distribution of fixation points. The main advantage of this index is its easy implementation in different areas. Studies have confirmed that the spatial distribution of fixations varies according to the mental workload [8, 10, 11]. The NNI was computed on epochs of 1 min [12] for each participant. Averaged NNI values were used as dependent variables in a repeated measures ANOVA design using two minute periods as repeated factor. Results showed a main effect of Time ($F_{11,143} = 4.41$, $p < .001$; Fig. 4). Duncan post-hoc testing showed that the visual strategy used in the first two minutes is significantly different from all other periods. Entropy rate. The entire visual area has been classified into two areas of interest (AOI), corresponding to the two images shown and the maximum number and duration of fixations made on each AOI were measured for each minute. Then, the entropy rate is introduced as a measurement of scan randomness for these AOI [5]. Entropy rate (H-rate) has the unit of bits/s (the information provided by a single observation, measured in bits on seconds). A higher H-rate value represents a more scan randomness pattern. In the present study, the entropy rate was calculated by using all scan paths performed by subjects. The entropy rates (H_rate) of the one-length sequences for the two used images were calculated as a measurement of scan randomness. Averaged H_rate values were used as dependent variables in a repeated measures ANOVA design using two minute periods as repeated factor. Results showed a main effect of Time ($F_{11,143} = 3.69$, $p < .001$). Duncan post hoc testing showed a greater regularity in visual exploration in the first two minutes, in line with the values of NNI.

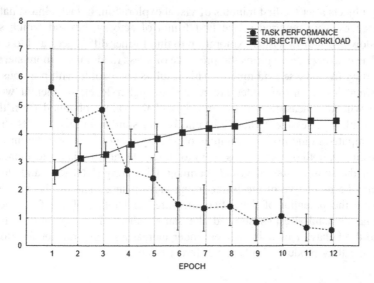

Fig. 3. Task performance (number of differences found) and subjective workload (ratings from 1 to 5) along time. Error bars denote .95 confidence intervals.

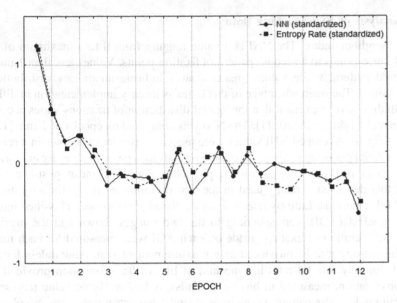

Fig. 4. NNI and Entropy rate by epoch. Measures have been standardized (z-scores) to plot them together.

4 Discussion and Conclusions

The main objective of this research was to compare two approaches to the analysis of scanpath that have been reported to be sensitive to variations in mental workload. Performance and subjective data showed a significant increase in the difficulty perceived by subjects after the first minutes of visual exploration. In fact, almost half of the total differences were reported in the first 6 min of activity, beyond which subjects engage a slow and narrow visual exploration to find further differences and to complete the task. The entropy rate confirms the presence of a less random and more stereotyped pattern starting from the second minute. This result is compatible with previous studies indicating that lower entropy rates are related to higher levels of mental workload. Consistent with the measurement of entropy and the values obtained from the self-reported workload assessment, the NNI trend is very similar. In fact, results show that the NNI computed in the first two minutes of activity is significantly different from the rest. Interestingly, both ocular indices seem to anticipate the drop in performance that is visible after the sixth minute of visual scanning. Clustering (NNI < 1) and stereotypy (H = 0) may be predictive of the forthcoming performance. This study was designed for assessing the potential of these two measures under the effect of an increasing visuo-spatial demand. Results showed the same pattern. Forthcoming studies may investigate the behavior of both indices under increasing temporal demand to understand whether they produce similar or different trends.

References

1. Cahill, J., et al.: Adaptive automation and the third pilot: managing teamwork and workload in an airline cockpit. In: Longo, L., Leva, M.C. (eds.) H-WORKLOAD 2017. CCIS, vol. 726, pp. 161–173. Springer, Cham (2017). https://doi.org/10.1007/978-3-319-61061-0_10
2. Muñoz-de-Escalona, E., Cañas, J.J.: Latency differences between mental workload measures in detecting workload changes. In: Longo, L., Leva, M.C. (eds.) H-WORKLOAD 2018. CCIS, vol. 1012, pp. 131–146. Springer, Cham (2019). https://doi.org/10.1007/978-3-030-14273-5_8
3. Peissl, S., Wickens, C., Baruah, R.: Eye-tracking measures in aviation: a selective literature review. Int. J. Aerosp. Psychology. **28**, 1–15 (2018). https://doi.org/10.1080/24721840.2018.1514978
4. Galpin, A.J., Underwood, G.: Eye movements during search and detection in comparative visual search. Percept. Psychophys. **67**(8), 1313–1331 (2005)
5. Tole, J.R., Stephens, A.T., Vivaudou, M., Ephrath, A.R., Young, L.R.: Visual scanning behavior and pilot workload (1983)
6. Harris, R.L., Glover, B.J., Spady, A.A.: Analytical Techniques of Pilot Scanning Behavior and Their Application (Report No. NASA TP-2525). NASA Langley Research Center, Hampton, VA, US (1986). https://ntrs.nasa.gov/archive/nasa/casi.ntrs.nasa.gov/19860018448.pdf
7. Di Nocera, F., Camilli, M., Terenzi, M.: A random glance at the flight deck: pilots' scanning strategies and the real-time assessment of mental workload. J. Cogn. Eng. Decis. Mak. **1**(3), 271–285 (2007)
8. Camilli, M., Nacchia, R., Terenzi, M., Di Nocera, F.: ASTEF: a simple tool for examining fixations. Behav. Res. Methods **40**(2), 373–382 (2008)
9. Tattersall, A.J., Foord, P.S.: An experimental evaluation of instantaneous self-assessment as a measure of workload. Ergonomics **39**(5), 740–748 (1996)
10. Dillard, M.B., et al.: The Sustained Attention to Response Task (SART) does not promote mindlessness during vigilance performance. Hum. Factors **56**(8), 1364–1379 (2014)
11. Fidopiastis, C.M., et al.: Impact of automation and task load on unmanned system operator's eye movement patterns. In: Schmorrow, Dylan D., Estabrooke, Ivy V., Grootjen, M. (eds.) FAC 2009. LNCS (LNAI), vol. 5638, pp. 229–238. Springer, Heidelberg (2009). https://doi.org/10.1007/978-3-642-02812-0_27
12. Di Nocera, F., Ranvaud, R., Pasquali, V.: Spatial pattern of eye fixations and evidence of ultradian rhythms in aircraft pilots. Aerosp. Med. Hum. Performance **86**(7), 647–651 (2015)

Eye-Tracking Metrics as an Indicator of Workload in Commercial Single-Pilot Operations

Anja K. Faulhaber[1(✉)] and Maik Friedrich[2]

[1] Institute of Flight Guidance, TU Braunschweig, Braunschweig, Germany
a.faulhaber@tu-braunschweig.de
[2] German Aerospace Center, Braunschweig, Germany

Abstract. There is a current trend in commercial aviation that points toward a possible transition from two-crew to single-pilot operations (SPO). The workload on the single pilot is expected to be a major issue for SPO. In order to find the best support solutions for the pilot in SPO, a thorough understanding of pilot workload is required. The present study aims at evaluating pilot workload by means of eye-tracking metrics. A flight simulator study was conducted with commercial pilots. Their task was to fly short approach and landing scenarios with or without the support of a second pilot. The results showed that fixation frequencies were higher during SPO, average dwell durations decreased, and participants transitioned more frequently between different areas of interest. These results suggest that particularly the temporal demand requires adequate support for a possible transition to SPO. The eye-tracking metrics support results obtained from subjective workload ratings.

Keywords: Single-pilot · Commercial aviation · Workload · Temporal demand · Eye tracking

1 Introduction

The flight deck in commercial aviation aircraft is built upon the principle of redundancy – all systems are available twice so that there is a backup in case of a failure. Accordingly, there are also two pilots in the cockpit who assume the roles of the pilot flying (PF) and the pilot monitoring (PM). The two pilots handle the workload throughout the flight together with the help of flight deck automation. Automated systems have in fact taken over many of the tasks originally accomplished by the pilots. This has led to such an advanced degree of autonomy that a transition from two-crew operations (TCO) to single-pilot operations (SPO) has become a topic of interest in commercial aviation. From a human-centered perspective, one of the major challenges in a possible introduction of SPO is workload [1]. Workload can be defined in terms of the "relationship between task demand and the relative capacity of a human being to perform the task effectively and within a given period of time" [2]. Studies have shown that high levels of workload can increase fatigue and lead to degraded performance (e.g., [3, 4]). It is therefore of particular interest to study workload during SPO

© Springer Nature Switzerland AG 2019
L. Longo and M. C. Leva (Eds.): H-WORKLOAD 2019, CCIS 1107, pp. 213–225, 2019.
https://doi.org/10.1007/978-3-030-32423-0_14

thoroughly to provide safe solutions for future concepts of operation. The literature review revealed three primary measures of workload: subjective ratings, performance-related measures, and physiological responses (e.g., [5, 6]). A preliminary analysis of performance data and subjective workload ratings from the present study indicated that pilots did not perceive workload as generally higher during SPO. Only the temporal demand was significantly increased. Moreover, workload was particularly high during abnormal events in SPO (for details see [7]). The objective of the present study is to investigate whether physiological responses in the form of eye-tracking data can serve as an objective measure to support and complement these preliminary results. Hence, this article focuses on the analysis of eye-tracking metrics as workload indicators.

1.1   Related Work

Workload is one of the major topics in the research field of commercial SPO. Many studies were conducted to assess various concepts of operation and new tools to support the pilot in SPO during high workload phases. Examples for the investigation of new SPO tools and systems are the operational alerting concept [8], augmented reality glasses [9], a virtual pilot assistant system [10], a cognitive pilot-aircraft interface [11], or a recommender system [12]. Additionally, advanced automation is frequently mentioned as a prerequisite to support the pilot in SPO and to reduce workload. In particular the importance of human-autonomy teaming has been suggested and a lot of research is being conducted in this area, too [13–16]. With regard to the concept of operation for SPO, there are several options that are being discussed (for an overview see [17, 18]). Most of these concepts include either advanced autonomy, a ground operator who can support the pilot when needed, or a combination of both. We found only few studies in the literature assessing workload during SPO in a present-day flight deck with no additional support for the pilot. A team of researchers at NASA studied workload and performance comparing TCO and SPO to the concept of reduced-crew operations in which one pilot is resting but can be called into the cockpit during high workload phases [19–24]. Their results suggest that workload is higher during SPO even though significance of the results is not reported clearly. Particularly during abnormal events the performance of pilots in SPO was poorer than during TCO. The studies mentioned previously used mostly subjective ratings and performance-related measures to assess workload in SPO. However, particularly subjective ratings are vulnerable to a variety of factors such as individual differences in personality traits and cognitive skills (e.g., [25, 26]). Moreover, when flying an aircraft, workload changes dynamically due to the varying task demand throughout the flight. Physiological responses might capture dynamic workload changes better than subjective ratings and measures of performance. In this context, eye-tracking metrics have frequently been proposed for workload estimation, particularly in relation to the field of aviation (for an overview see [27–29]). To the authors' knowledge, though, there is no study that reported eye-tracking metrics for workload estimation in the context of SPO so far. Using eye-tracking metrics as an objective measure to complement subjective workload ratings and performance measures might provide valuable insights and implications for possible future SPO in commercial aviation.

1.2   Present Study

The present study aims at identifying whether eye-tracking metrics may serve as an objective indicator of workload in SPO. The goal is to test neither a specific concept for SPO nor a new tool. Instead, we investigate what pilots can achieve without further support in a present-day cockpit. The results may be used to derive what kind of support solutions are in fact needed in future commercial SPO. An eye-tracking study was conducted in a fixed-base Airbus A320 flight simulator. Participants had to fly approach and landing scenarios manually with the Instrument Landing System (ILS). Approach and landing were chosen because workload is expected to reach critical levels particularly in these terminal area phases of flight [1]. Participants flew various scenarios with different workload levels both in a TCO condition with a PM and in a SPO condition without the PM. We chose eye-tracking metrics based on previous studies given that several metrics have been used in the context of workload assessment before. For example, average fixation duration and fixation frequency are commonly used to study workload but the empirical results are not conclusive (e.g., [30, 31]). Moreover, dwell-related analyses are frequently used, too. According to [32], a dwell is defined as one visit to an Area of Interest (AoI) from the moment of entering until the exit. It can therefore have its own duration and it can be composed of multiple fixations. Studies have shown that the average dwell duration increases with higher levels of mental workload and that this is due to a higher number of fixations per dwell (e.g., [33, 34]). Due to the absence of the PM, we furthermore expect that pilots need to manage and monitor more instruments and displays during SPO. This task load might be reflected in the transitions made between different AoIs. Therefore, we additionally investigate transition behavior in terms of transition frequency and average transition duration. In summary, we analyze the following eye-tracking metrics to assess workload during SPO: fixation frequency, average fixation duration, average dwell duration, average number of fixations per dwell, transition frequency, and average transition duration.

2   Methods

2.1   Participants

Fourteen pilots participated in the study. Two datasets had to be excluded from analysis due to technical issues. Moreover, two of the participants wore glasses which led to bad data quality so that their data could not be used either. In the end, data from ten participants served for analysis; all of them were male. Four of them were first officers and three of them captains. The remaining three had no rank because they were not working for an airline. Table 1 shows additional demographic information about the participants and their flying experience. All of the pilots participated voluntarily and did not receive any payment. The study was performed according to institutional and national standards for the protection of human subjects.

Table 1. Demographic information about the participants.

	Mean	SD	Range
Age	39.9	9.1	26–55
Total flight hours	4556	3699	300–10260
Airbus flight hours	984.1	1755.7	0–5000
Simulator hours	237.4	185.1	60–700

2.2 Design and Apparatus

The study was conducted in an Airbus A320 fixed-base simulator at the Institute of Flight Guidance, TU Braunschweig (Fig. 1). Participants had to fly short ILS approach and landing scenarios at Frankfurt Airport, runway 25 left. A 2 × 3 factorial within-subject design was used with the factors crew configuration (TCO and SPO) and scenario (baseline, turbulence, and abnormal). All scenarios located the aircraft at a starting position 8 nm from the runway at an altitude of about 2600 ft lined up for final approach and landing. The airspeed was set to 180 kt, the landing gear was still retracted and the flaps were already extended to 15° (indication 2).

Fig. 1. Cockpit of the Airbus A320 fixed-base flight simulator.

The three scenarios were designed to elicit different levels of workload. In the baseline scenario, workload was low. There were no clouds and the view was unrestricted. The wind was calm with a velocity of 3 kt. In the turbulence scenario, moderate turbulence was induced to increase the workload level slightly. In the abnormal scenario, an engine fire was simulated once the aircraft reached and altitude of 1800 ft. Such an engine fire first triggers a loud warning sound in the cockpit. Then, pilots are expected to perform an abnormal procedure as displayed in the electronic centralized aircraft monitor (ECAM). This procedure involves the following steps:

1. Pull the thrust lever of the affected engine to idle.
2. Shut down the affected engine via the master switch.
3. Push the respective engine fire button on the overhead panel.
4. After 10 s, discharge *Agent 1* to extinguish the fire.
5. Notify Air Traffic Control (ATC) about the emergency.
6. Discharge *Agent 2* if the engine is still on fire after 30 s, as indicated by the light of the engine fire button.

The abnormal scenario was, hence, designed to produce the highest workload due to an increase in the task load and a higher difficulty of flying with only one working engine. Eye-tracking data, simulator parameters, and video recordings were collected throughout the trials. Moreover, we applied the NASA Task Load Index (TLX) [35] to obtain subjective workload ratings. It consists of six workload subscales – mental demand, physical demand, temporal demand, performance, effort, and frustration – which are all rated on a scale from 0–100. Eye-tracking data and TLX workload ratings were collected only for the PF sitting in the left seat. The present paper focuses on the analysis of eye-tracking data as an objective measure to assess workload in SPO. Eye-tracking data were collected with SMI Eye Tracking Glasses (SensoMotoric Instruments, Germany). The glasses record gaze position, eye direction and head position binocularly with a sampling rate of 60 Hz.

2.3   Procedure

Two pilots participated in one session – *pilot A* and *pilot B*. They both participated in the experiment as PF and served as PM for each other in the respective TCO conditions. Upon arrival, they were first informed about the experiment and their task in a briefing session. The briefing contained information about the initial configuration of the aircraft and the weather conditions. Moreover, participants were informed that abnormal events such as system failures could occur but no further details were given. The participants received the required charts and checklists and gave informed consent. Afterward, depending on their prior experience, they both flew one or two training trials as PF. Once the training trials were finished, the eye tracker was calibrated in a one-point calibration and the experiment started. There were four experimental blocks. In the first block *pilot A* flew the three scenarios in the SPO condition. *Pilot B* was waiting in the briefing room and completed a demographic questionnaire. The experimenter was always in the cockpit, started each scenario by deactivating the initial freeze mode, and ended the scenario after touchdown. One scenario run lasted about 2.5 min and after it was finished, participants completed the NASA TLX and the simulator was reset to the initial configuration for the next scenario. When *pilot A* had finished all three scenarios in SPO, the second block started in which *pilot B* joined in as PM. They again flew the three scenarios together in the TCO condition. Afterward, they changed seats for the third block and *pilot B* flew the three scenarios first with *pilot A* as PM in the TCO condition. In the final fourth block, *pilot A* left the cockpit and *pilot B* flew the scenarios alone in the SPO condition. The order of the scenarios was counterbalanced to avoid an effect of order. When the experimental blocks were completed, debriefing interviews were conducted. The session took about two hours in total.

2.4   Data Analysis

The software Eye-Tracking Analyser by DLR [36] with a velocity-based fixation detection algorithm as described by [37] was used for the analysis of the eye-tracking data. Several AoIs were defined: (1) attitude indicator, (2) airspeed indicator, (3) altimeter (including variometer), (4) course indicator, (5) mode indicator, (6) navigation display (ND), (7) ECAM, (8) multifunction display (MFD), (9) gear lever, (10) multifunction control display unit (MCDU), (11) radio management panel, (12) thrust levers, (13) flap lever and speed brakes, (14) attention getter panel, (15) flight control unit (FCU), (16) overhead panel, (17) left window, (18) front window, (19) right window. Within these AoIs, only data obtained between an altitude of 2500 ft and touchdown were analyzed. Wilcoxon signed-rank tests were conducted for significance testing with an alpha-level of 0.05. Moreover, effect sizes r were calculated as described in [38].

3   Results

The preliminary data analysis of the subjective NASA TLX ratings had revealed that workload ratings were significantly higher in the abnormal scenario as compared to the baseline scenario. The turbulence scenario did not reach significantly higher workload ratings than the baseline scenario, though (for details see [7]). One participant even reported that he had not noticed the turbulence at all throughout the trials. The turbulence scenario was therefore excluded from our analyses of the eye-tracking metrics and only results from baseline and abnormal scenarios are reported in the following.

3.1   Fixations on Areas of Interest

The first analysis concerned the fixation frequency on all AoIs. Fixation frequency was defined as the number of fixations per minute. The analysis showed that participants in general fixated on AoIs more frequently in the SPO condition (Fig. 2, left). In the baseline scenario they made on average 77.504 fixations per minute (SD = 14.482) during TCO and 84.102 fixations per minute (SD = 15.782) during SPO. This difference was statistically significant ($z = -2.09$, $p = 0.037$, $r = 0.467$). The fixation frequency was even higher in the abnormal scenario both during TCO (M = 81.24, SD = 16.456) and SPO (M = 84.72, SD = 12.543) compared to the respective baseline scenarios. The difference in fixation frequency between TCO and SPO did not reach significance for the abnormal scenario, though ($z = -1.07$, $p = 0.322$, $r = 0.239$).

Second, the average fixation duration was analyzed. Results indicated that fixations were on average slightly shorter in the SPO condition as compared to the TCO condition (Fig. 2, right). However, the Wilcoxon signed-rank test did reach significance

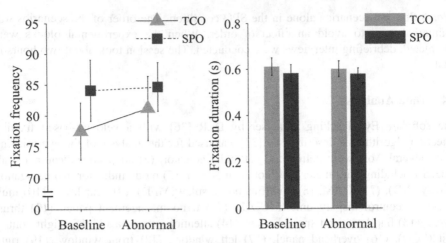

Fig. 2. Fixation frequency on all AoIs on the left and average fixation duration in seconds on the right. Error bars show the standard error of the mean.

neither for the baseline scenario (z = 0.866, p = 0.432, r = 0.194) nor for the abnormal scenario (z = 0.663, p = 0.557, r = 0.148). In summary, the results suggest that fixations were shorter and more frequent in the SPO condition.

3.2 Dwells on Areas of Interest

The average duration of dwells was analyzed and the results revealed that dwells on AoIs were on average shorter in duration in the SPO condition (Fig. 3, left). This difference reached statistical significance in the baseline scenario with means of 0.806 s (SD = 0.217) for TCO and 0.72 s (SD = 0.17) for SPO (z = 2.192, p = 0.027, r = 0.49). The average dwell duration in the SPO abnormal scenario (M = 0.693, SD = 0.101) was also shorter than in the TCO abnormal scenario (M = 0.774, SD = 0.168) but the effect was not significant (z = 1.784, p = 0.084, r = 0.399). These results also indicate that the average dwell durations were shorter in the abnormal scenario as compared to the baseline scenario for both TCO and SPO conditions. Moreover, the average number of fixations per dwell was analyzed. The aim of this analysis was to find out whether the dwell durations in the SPO condition were shorter only due to the shorter fixation durations or whether there were also fewer fixations per dwell. The results revealed that there were indeed fewer fixations per dwell in the SPO condition (Fig. 3, right). The results reached significance in the baseline scenario (z = 2.191, p = 0.027, r = 0.49), where the TCO condition reached a mean of 1.312 fixations per dwell (SD = 0.171) and the SPO condition 1.228 fixations per dwell (SD = 0.106). In the abnormal scenario, the difference in means between TCO (M = 1.3, SD = 0.237) and SPO (M = 1.194, SD = 0.05) did not reach statistical significance (z = 1.172, p = 0.275, r = 0.262). In conclusion, the dwell duration was generally shorter in the SPO condition. This effect was due to shorter as well as fewer fixations per dwell.

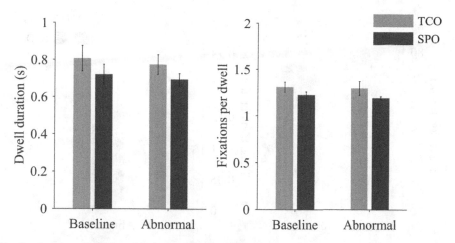

Fig. 3. Average dwell duration for all AoIs on the left and average number of fixations per dwell on the right. Error bars show the standard error of the mean.

3.3   Transitions Between Areas of Interest

Transitions between AoIs were analyzed; more precisely, one of the measures used was the transition frequency. In accordance with the fixation frequency, the transition frequency was defined as the number of transitions between AoIs per minute. The results showed that participants transitioned more frequently between AoIs in the SPO condition (Fig. 4, left). In the TCO baseline scenario participants transitioned on average 60.313 times per minute (SD = 16.816) between AoIs. In the SPO condition, the average transition frequency increased to 69.214 transitions per minute (SD = 17.096). The Wilcoxon signed-rank test showed that this difference was statistically significant ($z = -2.497$, $p = 0.01$, $r = 0.558$). Similar to the fixation frequency, the results for the transition frequency in the abnormal scenario were higher as compared to the respective baseline scenario conditions. The means for the TCO abnormal scenario (M = 64.639, SD = 18.421) and the SPO abnormal scenario (M = 70.59, SD = 10.552) did not differ significantly ($z = -1.07$, $p = 0.322$, $r = 0.239$). The final analysis presented in this paper concerns the average transition duration. The results indicated that transitions were on average shorter in duration in the SPO condition as compared to the TCO condition (Fig. 4, right). Transitions took on average 0.177 s (SD = 0.133) in the TCO baseline scenario and 0.169 s (SD = 0.147) in the TCO abnormal scenario. In the SPO condition, transition durations decreased to 0.162 s (SD = 0.081) in the baseline scenario and to 0.137 s (SD = 0.04) in the abnormal scenario. The Wilcoxon signed-rank test did not reach significance either for the baseline scenario ($z = -0.561$, $p = 0.625$, $r = 0.125$) or for the abnormal scenario ($z = -0.051$, $p = 1$, $r = 0.011$). Nevertheless, it can be summarized that participants transitioned more frequently between AoIs in the SPO condition and that transitions were on average slightly shorter in duration.

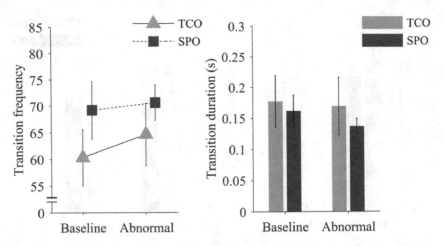

Fig. 4. Transition frequency between all AoIs on the left and average transition duration in seconds on the right. Error bars show the standard error of the mean.

4 Discussion and Conclusion

To sum up, the results showed that in the SPO condition, fixation frequencies were higher and average fixation durations were shorter. Dwells were also on average shorter in duration and composed of fewer fixations in the SPO condition. Participants transitioned more frequently between AoIs and the transition duration was on average shorter during SPO. The differences in means between TCO and SPO for the fixation frequency, the average dwell duration, and the transition frequency were significant in the baseline scenario.

The present study has certain limitations that need to be discussed. The number of participants was relatively small. A further study with more participants should be conducted to confirm the results. Only data of the PF, who was always flying from the left seat, was collected. Almost half of the participants were first officers used to flying from the right seat. Some participants mentioned in this context that flying with the left hand might have increased their perceived workload. Additionally, not all the participants were experienced on the A320 and a lack of familiarity with the aircraft type might have affected pilot workload, too. Moreover, communication with ATC is a major aspect of workload that could not be simulated with high fidelity in the present study. Finally, in the abnormal scenario, not all of the participants performed the engine fire procedure. Only three participants performed the procedure in both TCO and SPO conditions. Five participants disregarded the procedure in both conditions because they felt it was safer to handle the fire on the ground. The remaining two participants performed the procedure only in the TCO condition and disregarded it in the SPO condition. One of them stated in the debriefing that he had therefore perceived workload in the TCO abnormal scenario as higher than in the SPO condition. This might have distorted the results in the abnormal scenario and led to the fact that the results were not statistically significant. Nevertheless, the results serve as an indicator for workload differences between TCO and SPO. More precisely, the descriptive statistics for both the baseline and the abnormal scenario showed shorter average fixation and dwell durations during SPO. Interestingly, these results are contrary to the findings from other studies which associated higher mental workload with longer fixation and dwell durations (e.g., [33, 34]). Combined with the higher frequency in fixations and transitions, our results suggest that the participants' eye movements were faster during SPO and reflected stress, time pressure, and hence a higher temporal demand. Thus, workload was not increased in terms of a higher mental demand but our results particularly support an increase in the temporal demand. The findings regarding the higher temporal demand during SPO support our results from the NASA TLX which had already revealed that the ratings on the temporal demand subscale were significantly higher for SPO compared to TCO [7]. The results are also in line with previous studies which identified an additional task load on the pilot due to the absence of the PM in SPO (e.g., [39]). This additional task load can explain the increase in the temporal demand on the pilot during SPO. Further studies have already investigated how the task load can be reduced by means of different concepts of operation for SPO [40, 41]. Automation plays an important role here but also the concept of a human ground operator who supports pilots during high workload phases has gained popularity and empirical studies provided promising results [42]. However, further research is still required in this area. To sum up, the present results are only to some degree in line with previously reported results concerning pilot workload in SPO with present-day flight deck design and no additional support [19]. In the present study, workload was not found to be generally higher during SPO as opposed to TCO. However, the present results can be interpreted in terms of an increased temporal demand. Data from both subjective workload ratings and objective eye-tracking metrics suggest that temporal demand is higher during SPO than TCO. Future SPO concepts should take these results into account by providing support to reduce the temporal demand. Further research is required to determine how optimal support solutions should be designed to support the single pilot effectively. On a more general note, our results suggest that eye-tracking metrics are sensitive to differences in temporal demand. This is particularly interesting in light of the latest research regarding online pilot performance monitoring during flight via eye-tracking [43]. Further research is required to evaluate the use of eye-tracking metrics for online workload monitoring.

Acknowledgments. This research was conducted within the graduate program "Gendered configurations of humans and machines. Interdisciplinary analyses of technology" (KoMMa.G) funded by the federal state of Lower Saxony, Germany. The authors would like to thank Prof. Dr. Peter Hecker, Dr. Thomas Feuerle, Stefan Seydel, and Tatjana Kapol for their support throughout the project.

References

1. Koltz, M.T., et al.: An investigation of the harbor pilot concept for single pilot operations. Procedia Manuf. **3**, 2937–2944 (2015)

2. Fernández Medina, K.: Measuring the mental workload of operators of highly automated vehicles. In: Longo, L., Leva, M.C. (eds.) H-WORKLOAD 2018. CCIS, vol. 1012, pp. 112–128. Springer, Cham (2019). https://doi.org/10.1007/978-3-030-14273-5_7
3. Fan, J., Smith, Andrew P.: The impact of workload and fatigue on performance. In: Longo, L., Leva, M.C. (eds.) H-WORKLOAD 2017. CCIS, vol. 726, pp. 90–105. Springer, Cham (2017). https://doi.org/10.1007/978-3-319-61061-0_6
4. Morris, C.H., Leung, Y.K.: Pilot mental workload. How well do pilots really perform? Ergonomics **49**, 1581–1596 (2006)
5. Hancock, P.A.: Whither workload? Mapping a path for its future development. In: Longo, L., Leva, M.C. (eds.) H-WORKLOAD 2017. CCIS, vol. 726, pp. 3–17. Springer, Cham (2017). https://doi.org/10.1007/978-3-319-61061-0_1
6. Hancock, P.A., Meshkati, N. (eds.): Human Mental Workload. North Holland Press, Amsterdam (1988)
7. Faulhaber, A.K.: From crewed to single-pilot operations. Pilot performance and workload management. In: Proceedings of the 20th International Symposium on Aviation Psychology, Dayton, OH (2019)
8. Reitsma, J.P., van Paassen, M.M., Mulder, M.: Operational alerting concept for commercial single pilot operations systems. In: Proceedings of the 20th International Symposium on Aviation Psychology, Dayton, OH (2019)
9. Tran, T.H., Behrend, F., Fünning, N., Arango, A.: Single pilot operations with AR-glasses using Microsoft HoloLens. In: IEEE/AIAA 37th Digital Avionics Systems Conference, London, UK (2018)
10. Lim, Y., Ramasamy, S., Gardi, A., Sabatini, R.: A virtual pilot assistant system for single pilot operations of commercial transport aircraft. In: 17th Australian International Aerospace Congress: AIAC 2017, pp. 139–145. Engineers Australia, Royal Aeronautical Society, Barton (2017)
11. Liu, J., Gardi, A., Ramasamy, S., Lim, Y., Sabatini, R.: Cognitive pilot-aircraft interface for single-pilot operations. Knowl. Based Syst. **112**, 37–53 (2012)
12. Dao, A.-Q.V., et al.: Evaluation of a recommender system for single pilot operations. Procedia Manuf. **3**, 3070–3077 (2015)
13. Matessa, M.: Using a crew resource management framework to develop human-autonomy teaming measures. In: Baldwin, C. (ed.) AHFE 2017. AISC, vol. 586, pp. 46–57. Springer, Cham (2018). https://doi.org/10.1007/978-3-319-60642-2_5
14. Cover, M., Reichlen, C., Matessa, M., Schnell, T.: Analysis of airline pilots subjective feedback to human autonomy teaming in a reduced crew environment. In: Yamamoto, S., Mori, H. (eds.) HIMI 2018. LNCS, vol. 10905, pp. 359–368. Springer, Cham (2018). https://doi.org/10.1007/978-3-319-92046-7_31
15. Shively, Robert J., Lachter, J., Koteskey, R., Brandt, Summer L.: Crew resource management for automated teammates (CRM-A). In: Harris, D. (ed.) EPCE 2018. LNCS (LNAI), vol. 10906, pp. 215–229. Springer, Cham (2018). https://doi.org/10.1007/978-3-319-91122-9_19
16. Shively, R.J., Brandt, S.L., Lachter, J., Matessa, M., Sadler, G., Battiste, H.: Application of human-autonomy teaming (HAT) patterns to reduced crew operations (RCO). In: Harris, D. (ed.) EPCE 2016. LNCS (LNAI), vol. 9736, pp. 244–255. Springer, Cham (2016). https://doi.org/10.1007/978-3-319-40030-3_25
17. Neis, S.M., Klingauf, U., Schiefele, J.: Classification and review of conceptual frameworks for commercial single pilot operations. In: IEEE/AIAA 37th Digital Avionics Systems Conference, London, UK (2018)
18. Vu, K.-P.L., Lachter, J., Battiste, V., Strybel, T.Z.: Single pilot operations in domestic commercial aviation. Hum. Fact. **60**, 755–762 (2018)

19. Bailey, R.E., Kramer, L.J., Kennedy, K.D., Stephens, C.L., Etherington, T.J.: An assessment of reduced crew and single pilot operations in commercial transport aircraft operation. In: IEEE/AIAA 36th Digital Avionics Systems Conference, St. Petersburg, FL (2017)
20. Etherington, T.J., Kramer, L.J., Bailey, R.E., Kennedy, K.D., Stephens, C.L.: Quantifying pilot contribution to flight safety for normal and non-normal airline operations. In: IEEE/AIAA 35th Digital Avionics Systems Conference, Sacramento, CA (2016)
21. Etherington, T.J., Kramer, L.J., Kennedy, K.D., Bailey, R.E., Last, M.C.: Quantifying pilot contribution to flight safety during dual generator failure. In: IEEE/AIAA 36th Digital Avionics Systems Conference, St. Petersburg, FL (2017)
22. Etherington, T.J., Kramer, L.J., Bailey, R.E., Kennedy, K.D.: Quantifying pilot contribution to flight safety during an in-flight airspeed failure. In: Proceedings of the 19th International Symposium on Aviation Psychology, Dayton, OH (2017)
23. Kramer, L.J., Etherington, T.J., Last, M.C., Bailey, R.E., Kennedy, K.D.: Quantifying pilot contribution to flight safety during drive shaft failure. In: IEEE/AIAA 36th Digital Avionics Systems Conference, St. Petersburg, FL (2017)
24. Kramer, L.J., Etherington, T.J., Bailey, R.E., Kennedy, K.D.: Quantifying pilot contribution to flight safety during hydraulic systems failure. In: Stanton, N.A. (ed.) AHFE 2017. AISC, vol. 597, pp. 15–26. Springer, Cham (2018). https://doi.org/10.1007/978-3-319-60441-1_2
25. Guastello, S.J., Marra, D.E., Correro, A.N., Michels, M., Schimmel, H.: Elasticity and rigidity constructs and ratings of subjective workload for individuals and groups. In: Longo, L., Leva, M.C. (eds.) H-WORKLOAD 2017. CCIS, vol. 726, pp. 51–76. Springer, Cham (2017). https://doi.org/10.1007/978-3-319-61061-0_4
26. Guastello, S.J., Shircel, A., Malon, M., Timm, P.: Individual differences in the experience of cognitive workload. Theor. Issues Ergon. Sci. 16, 20–52 (2013)
27. Peißl, S., Wickens, C.D., Baruah, R.: Eye-tracking measures in aviation. A selective literature review. Int. J. Aerosp. Psychol. 28, 98–112 (2018)
28. Glaholt, M.G.: Eye tracking in the cockpit. A review of the relationships between eye movements and the aviator's cognitive state. Defence Research and Development, Toronto, Canada (2014)
29. Ziv, G.: Gaze behavior and visual attention. A review of eye tracking studies in aviation. Int. J. Aviat. Psychol. 26, 75–104 (2017)
30. Itoh, Y., Hayashi, Y., Tsukui, I., Saito, S.: The ergonomic evaluation of eye movement and mental workload in aircraft pilots. Ergonomics 33, 719–733 (1990)
31. Ellis, K.K.E.: Eye tracking metrics for workload estimation in flight deck operations. Iowa City (2009)
32. Holmqvist, K., Nyström, M., Andersson, R., Dewhurst, R., Jarodzka, H., van de Weijer, J.: Eye Tracking. A Comprehensive Guide to Methods and Measures. Oxford University Press, Oxford (2015)
33. Tole, J.R., Stephens, A.T., Vivaudou, M., Ephrath, A.R., Young, L.R.: Visual scanning behavior and pilot workload. NASA, Washington, D.C. (1983)
34. Harris, R.L., Glover, B.J., Spady, A.A.: Analytical techniques of pilot scanning behavior and their application. NASA, Hampton, VA (1986)
35. Hart, S.G., Staveland, L.E.: Development of NASA-TLX (Task Load Index). Results of empirical and theoretical research. In: Hancock, P.A., Meshkati, N. (eds.) Human mental workload, pp. 139–183. North Holland Press, Amsterdam (1988)
36. Friedrich, M., Rußwinkel, N., Möhlenbrink, C.: A guideline for integrating dynamic areas of interests in existing set-up for capturing eye movement. Looking at moving aircraft. Behav. Res. Methods 49, 822–834 (2017)

37. Salvucci, D.D., Goldberg, J.H.: Identifying fixations and saccades in eye-tracking protocols. In: Proceedings of the Eye Tracking Research and Applications Symposium, pp. 71–78. ACM Press, New York (2000)
38. Field, A.: Discovering Statistics Using SPSS. Sage, Los Angeles (2009)
39. Schutte, P.C.: Task analysis of two crew operations in the flight deck. Investigating the feasibility of using single pilot. In: Proceedings of the 19th International Symposium on Aviation Psychology, Dayton, OH (2017)
40. Wolter, C.A., Gore, B.F.: A validated task analysis of the single pilot operations concept. NASA, Moffett Field, CA (2015)
41. Gore, B.F., Wolter, C.: A task analytic process to define future concepts in aviation. In: Duffy, V.G. (ed.) DHM 2014. LNCS, vol. 8529, pp. 236–246. Springer, Cham (2014). https://doi.org/10.1007/978-3-319-07725-3_23
42. Lachter, J., Brandt, S.L., Battiste, V., Matessa, M., Johnson, W.W.: Enhanced ground support. Lessons from work on reduced crew operations. Cogn. Technol. Work **19**, 279–288 (2017)
43. Dehais, F., Causse, M., Pastor, J.: Embedded eye tracker in a real aircraft. New perspectives on pilot/aircraft interaction monitoring. In: Proceedings from the 3rd International Conference on Research in Air Transportation, Fairfax, VA (2008)

EEG-Based Mental Workload and Perception-Reaction Time of the Drivers While Using Adaptive Cruise Control

Ennia Acerra[1], Margherita Pazzini[1], Navid Ghasemi[1],
Valeria Vignali[1], Claudio Lantieri[1(✉)], Andrea Simone[1],
Gianluca Di Flumeri[2,3], Pietro Aricò[2,3], Gianluca Borghini[2,3],
Nicolina Sciaraffa[2,4], Paola Lanzi[5], and Fabio Babiloni[2,3]

[1] Department of Civil, Chemical, Environmental and Materials Engineering
(DICAM), University of Bologna, Viale Risorgimento, 2, 40136 Bologna, Italy
{ennia.acerra2,margherita.pazzini2,navid.ghasemi3,
valeria.vignali,claudio.lantieri2,
andrea.simone}@unibo.it
[2] BrainSigns srl, via Sesto Celere, 00152 Rome, Italy
{gianluca.diflumeri,pietro.arico,gianluca.borghini,
nicolina.sciaraffa,fabio.babiloni}@uniromal.it
[3] Department of Molecular Medicine, Sapienza University of Rome, Rome, Italy
[4] SAIMLAL, Sapienza University of Rome, Rome, Italy
[5] DeepBlue srl, Piazza Buenos Aires 20, 00185 Rome, Italy
paola.lanzi@dblue.it

Abstract. Car driving is a complex activity, consisting of an integrated multi-task behavior and requiring different interrelated skills. Over the last years, the number of Advanced Driver Assistance systems integrated into cars has grown exponentially. So it is very important to evaluate the interaction between these devices and drivers in order to study if they can represent an additional source of driving-related distraction. In this study, 22 subjects have been involved in a real driving experiment, aimed to investigate the effect of the use of the Adaptive Cruise Control (ACC) on mental workload and Perception-Reaction Time of the drivers. During the test physiological data, in terms of brain activity through Electroencephalographic technique and eye gaze through Eye-Tracking devices, and vehicle trajectory data, through a satellite device mounted on the car, have been recorded. The results obtained show that the use of ACC caused an increase in mental workload and Perception-Reaction Time of the drivers.

Keywords: Electroencephalography · Eye-Tracking · Mental workload ·
Human factor · Car driving · Road safety · Adaptive Cruise Control ·
Perception-Reaction Time

© Springer Nature Switzerland AG 2019
L. Longo and M. C. Leva (Eds.): H-WORKLOAD 2019, CCIS 1107, pp. 226–239, 2019.
https://doi.org/10.1007/978-3-030-32423-0_15

1 Introduction

The Global status report on road safety 2018, launched by the World Health Organization (WHO), highlights that the number of annual road traffic deaths has reached 1.35 million. Road traffic injuries may now be considered as the leading causes of death among people aged 5–29 years [1]. Car driving is a complex activity, consisting in an integrated multi-task behavior engaging several processes and requiring different interrelated skills that rely on interconnected visual, motor and cognitive brain systems [2]. While driving, the interactions between the driver, the vehicle and the environment are continuous and numerous [3, 4]. Driver's common errors are largely correlated to overload, distractions, tiredness, or the simultaneous realization of other activities while driving. Secondary task distraction is a contributing factor in up to 23% of crashes and near-crashes [5, 6]. Considering the driver's error resulting in severe consequences in road transportations, the development of countermeasures to mitigate the human errors through training and technology and a better road system becomes critical [7]. Advanced driver assistance systems (ADAS) and Passive Safety Systems (PSS) (e.g. seatbelts, airbags) are two approaches used in modern vehicles to mitigate the risk of accidents or casualties resulting from human error. ADAS are electronic control systems that aid drivers while driving. They are designed to improve driver, passenger and pedestrian safety by reducing both the severity and the overall number of motor vehicle accidents. ADAS can warn drivers of potential dangers, intervene to help the driver remain in control in order to prevent an accident and, if necessary, reduce the severity of an accident in case if it cannot be avoided. Adaptive cruise control (ACC) is an ADAS that automatically adjusts the vehicle speed to maintain a safe distance from vehicles ahead. A vehicle equipped with ACC will thus reduce speed automatically, within limits, to match the speed of a slower vehicle that is following. ACC automates the operational control of headway and speed. It should reduce driver stress and human errors as a result of freeing up visual, cognitive and physical resources [8], and the number of hard accelerations and decelerations, encourage speed harmonization between vehicles and enable better merging behaviors [9].

Despite the potential benefits of ACC, negative behavioral adaptation (BA) may occur with its introduction [10]. Driver may use any freed visual, cognitive and physical resources to engage in non-driving tasks that he perceives as improving his productivity. However, these tasks may reduce his vigilance and attention to the primary driving task, which could result in driver distraction, and a failure to detect and respond to critical driving situations [11, 12]. When using ACC, drivers are more likely to perform in-vehicle tasks that they would not normally do [13], and their performance on a secondary task improves [14]. The visual demand of drivers decreases when they use ACC, allowing them to pay less attention the road ahead [15]. The main objective of this study was to investigate whether ACC can induce BA in drivers. 22 subjects have been involved in a real driving experiment, aimed to investigate the effect of the use of the Adaptive Cruise Control (ACC) on mental workload and Perception-Reaction Time of the drivers. Several neurophysiological studies about drivers' behaviors, in fact, have shown that the same experimental tasks are perceived differently, in terms of mental workload, if performed in a simulator or in a real environment [16]. Pierre et al. and Tong et al., moreover, have confirmed that not only the task perception, but also the driver behavior itself related to a specific condition, could change if the same condition is reproduced in simulators or in a real scenario [17, 18]. Therefore, the first key aspect of the present work is the real urban context employed to perform the experiments.

During the experiments, different parameters have been monitored:

- the drivers' mental workload, through objective measure by EEG technique and subjective measure through the NASA Task Load Index (NASA-TLX) questionnaire;
- the drivers' Perception-Reaction Time, through Eye-Tracking (ET) device;
- the vehicle trajectory and velocity, using a GPS device mounted on the car.

Electroencephalographic technique (EEG) has been demonstrated to be one of the techniques to infer, in real time, the mental workload experienced by the user, since it is a direct measure of brain activations and it is characterized by high temporal resolution, limited cost and invasiveness [19–21]. Also, it provides direct access to what is happening within human mind, thus providing objective assessment of cognitive phenomena [22]. Numerous research papers have demonstrated that braking behaviour and Perception-Reaction Time are useful indicators of the amount of attention a driver is allocating to the driving task. Especially for ACC, if drivers devote more resources to non-driving tasks when using ACC, they may show an impaired ability to respond to safety-relevant driving situations that require them to apply the brakes themselves (such as a lead vehicle braking suddenly or an ACC system failure). In response to a braking lead vehicle, drivers begin braking later when using ACC showing higher Perception-Reaction Time [15, 23]. In this paper, 22 subjects have been involved in a real driving experiment performed along the Tangenziale of Bologna (Italy). Each subject, after a proper experimental briefing, had to repeat the circuit two times within the same day and, during the second lap, three different events with both ACC enabled and disabled (respectively ON and Off conditions) have been acted, by involving a confederate vehicle along the route.

2 Methods

2.1 Sample

Twenty-two males drivers (mean age = 47.12 years $\pm$ 5.58, range: 35 $\div$ 55) took part in the study. They were selected in order to have a homogeneous experimental group in terms of age, sex and driving expertise. Everyone had normal vision and none of them wore eyeglasses or lenses, to avoid artefacts in eye-movement monitoring. They were paid and they did not know anything about the aims of the study in order to avoid any bias of their behavior. They only knew that the study aimed to test the mobile eye-recording device while driving. All subjects had a Category B driving license (for cars). None of them had previously driven on the road segment considered in this study.

Subjects had previously experience in driving a vehicle with automatic transmission and Adaptive Cruise Control (mean number of hours of experience with ACC = 3.31 years ± 1.81, range: 1 ÷ 5).

The experiment was conducted following the principles outlined in the Declaration of Helsinki of 1975, as revised in 2000. The study was approved by the Ethic Committee of University of Bologna. Informed consent and authorization to use the video graphical material were obtained from each subject on paper, after the explanation of the study.

2.2   Experimental Site

The subjects had to drive along the Tangenziale of Bologna (Italy), a bypass road, coplanar with the urban section of the A14 highway. It is a primary road, mainly straight with wide radius curves, with two lanes in each direction (excluding the emergency lane), with intersections at ground level only in correspondence with the junctions. This road has been chosen because it has right requirements for the application of the Adaptive Cruise Control as it allows speed higher than 60 km/h and it has a multi-lane carriageway in each direction with a horizontal signs in a god maintenance state.

Moreover, the route consisted in two laps of a "circuit" about 10 km long (Fig. 1).

Fig. 1. The circuit of the real driving experiment.

Each subject, after a proper experimental briefing, had to repeat the circuit two times within the same day. The first lap was considered as an "adaptation lap", because it was useful for the driver adaptation to the route, the vehicle and to the ACC system. During this lap the driver was free to experience the ACC system. The data recorded during the second lap were taken into account for the analysis. During this lap, the user drove half of the track with ACC enabled (ON condition) and other half with ACC disabled (OFF condition). The order of ACC on and off conditions had been randomized among the subjects, in order to avoid any order effect. The same car was used for the experiments, i.e. a Volkswagen Touareg, with diesel engine and automatic transmission. It was equipped with Adaptive Cruise Control (ACC). Finally, during the test lap (the 2nd one) three similar events for both ACC on and off conditions were simulated, by involving a confederate vehicle along the route: a car entering the traffic flow ahead of the experimental subject and braking in order to simulate a critical event (Fig. 2). The event type was selected as the most probable one coherently with the ACC mode of operation, as well as the safest to be acted, without introducing any risk for the actors, for the experimental subjects and for the traffic in general.

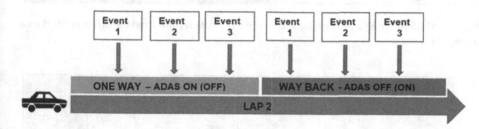

Fig. 2. Simulation of the events.

2.3   Procedure and Measurements

Subjects drove a route of 5 km + 5 km from along the Tangenziale of Bologna. They did not know the route in advance and they were given instructions about how and when to come back. Data collection started at 9 a.m. and finished at 18 p.m. on two different days, always in autumn, in a period with good meteorological conditions. All the subjects drove the same car. The vehicle was equipped with a Video Vbox Pro data recorder (Fig. 3). Two cameras and a GPS antenna were placed on the top of the car and connected to the Vbox data recorder. The complete driver's scene was recorded, including synchronized data on acceleration, speed and GPS coordinates. The system recorded speed (accuracy: 0.1 km/h), acceleration (1% accuracy), and distance with a 20 Hz sample rate. In order to evaluate the drivers' Perception-Reaction Times, an ASL Mobile Eye-XG device recorded the driver's eye movements (Fig. 3). The device consisted of two cameras attached to eyeglasses, one recording the right eye movements, and the second one recording the visual scene. In order not to obscure the normal field of view of the drivers, a mirror capable of reflecting the infrared light was installed in the eye camera recording the activity of the right eye. As already tested in Costa et al., the sampling rate for the eye-movement recording was 30 Hz with an accuracy of 0.5–1° (approximating the angular width of the fovea) [24–26].

A video for each subject was created using the ASL software with a cross superimposed to the scene showing the eye fixations. This allows researchers to detect the sequence of points of the scene fixated by the driver (Fig. 4).

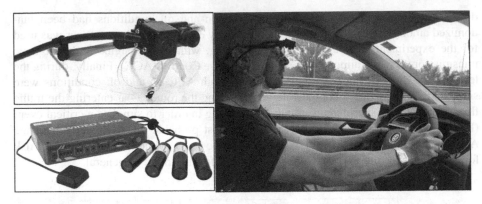

Fig. 3. ASL Mobile Eye-XG device, Video Vbox Pro data recorder and the driver inside the car during the test.

Fig. 4. Video created using the ASL software with a cross superimposed to the scene showing the eye fixations. In the left figure the vehicle was seen, in the right one it was not seen.

In order to get a good accuracy of the eye-movement recorder, a calibration procedure was carried out for each subject according to his eye status. The ASL Mobile Eye-XG and the Video VBOX PRO devices were installed in the back seat of the car, monitored by one of the researchers, who were asked not to talk to the driver, with apart from giving instructions about the direction and assistance in case of necessity. In order to evaluate the drivers' mental workload, subjective and objective measures, respectively through the EEG technique and the NASA Task Load Index (NASA-TLX) questionnaire, were adopted. In fact, it is widely accepted in scientific literature the limit of using subjective measures alone, such as questionnaires and interview, because of their intrinsic subjective nature and the impossibility to catch the "unconscious" phenomena behind human behaviors [22, 27]. The EEG signals have been recorded using the digital monitoring BEmicro system (EBNeuro, Italy). Twelve EEG channels (FPz, AF3, AF4, F3, Fz, F4, P3, P7, Pz, P4, P8 and POz), placed according to the 10–20 International System, were collected with a sampling frequency of 256 Hz, all referenced to both the earlobes, grounded to the Cz site, and with the impedances kept below 20 kΩ. The EEG data have been used to assess the mental workload of each driver through an innovative machine-learning algorithm developed [28] and validated in a previous driving study [5]. In addition, subjective measures of perceived mental workload have been collected from the subjects after both the tasks through the NASA Task Load indeX (NASA-TLX) questionnaire [5].

2.4 Data Analysis

For each braking event performed during the test lap (the 2nd one), for both ACC on and off conditions, EEG-based mental workload assessment and Perception-Reaction Time of the drivers were analyzed. The average value for each condition (ACC on and ACC off) was taken into consideration. The acquired EEG signals were digitally band-pass filtered by a 5th order Butterworth filter [1 ÷ 30 Hz]. The eyeblink artifacts were removed from the EEG using the REBLINCA method [29]. The EEG signal from the remaining eleven electrodes was then segmented in 2 second-epochs, shifted of 0.125 s, with the aim to have both a high number of observations in comparison with the number of variables, and to respect the condition of stationarity of the EEG signal [30]. For other sources of artifacts, specific procedures of the EEGLAB toolbox were applied, in order to remove EEG epochs marked as "artifact". The Power Spectral Density (PSD) has then been estimated by using the Fast Fourier Transform (FFT) in the EEG frequency bands defined for each subject by the estimation of the Individual Alpha Frequency (IAF) value [31]. In this regard, before starting with the experiment, the brain activity of each subject during a minute of rest (closed eyes) was recorded, in order to calculate the IAF. Thus, the Theta rhythms [IAF−6 ÷ IAF−2] over the Frontal sites and the Alpha rhythms [IAF−2 ÷ IAF+2] over the Parietal sites were investigated, because of their strict relationship with mental workload [32], and used to compute the mental Workload index (WL index). As introduced before, the WL index was calculated by using the machine learning approach proposed by Aricò and colleagues [28], the automatic stop-StepWise Linear Discriminant Analysis (asSWLDA) classifier. The time resolution of the provided WL index has been fixed at 8 s, since this value has been demonstrated to be a good trade-off between the resolution and the accuracy of the measure [19]. The Perception-Reaction Time (PRT), as defined by Olson and Sivak, is "the time from the first sighting of an obstacle until the driver applies the brakes" [33, 34]. It has been evaluated as the difference between the Reaction time and the Perception time.

The Perception Time is the time when the driver sees the braking of the confederate vehicle. The Reaction Time, instead, is the time when the driver starts to brake his vehicle. In order to calculate PRT value for each critical event performed during the test lap, the Video Vbox Pro data recorder and the ASL Mobile Eye-XG device were synchronized. The synchronization of the speed data and the eye-tracking data was obtained by the methodology used in Costa, Bonetti et al., Costa, Bichicchi et al. and in Costa, Simone et al. [35–38]. The Reaction Time was evaluated by Video Vbox-Pro output video, as the time in which the driver starts to brake, after having looked at the stop light of confederate vehicle (Fig. 5). The Perception Time was evaluated by ASL Mobile Eye-XG device, as the frame in which the confederate vehicle braked, its led stop became red and the driver saw the stop light. This represents the time of the first-fixation of the red stop light of the confederate vehicle (Fig. 5).

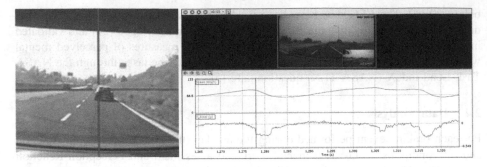

Fig. 5. First-fixation of the red led stop of the confederate vehicle and time of breaking. (Color figure online)

In Fig. 6 it was possible to highlight the range, corresponding to the Perception-Reaction Time of one braking event.

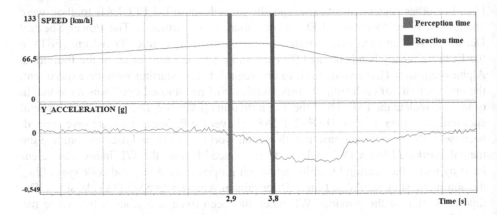

Fig. 6. Evaluation of the Perception-Reaction Time.

In order to evaluate the Perception Time, the ASL Mobile Eye-XG videos were analyzed frame-by-frame. To avoid the inclusion of saccadic movements, an element was considered as fixated when it was fixed for a minimum duration of two frames (66 ms). The threshold of 66 ms, which is low in comparison to a common filtering of 100 ms or higher usually found in eye-tracking studies, was dictated by the specific setting of this study that involved the recording of eye movements while driving. Lantieri et al. and Vignali et al. reported that in real traffic situations, in a high dynamical context of road driving, the duration of fixation is much lower than in other contexts and experimental settings. In a real driving setting with a dynamical visual scene, as in the case of the present study, rapid fixations may occur [39–41]. Also, at the end of task the subjects had to evaluate the experienced workload by filling the NASA-TLX questionnaire. Each subject has filled two questionnaires: one for the half of the track with ACC enabled (ON condition) and other half with ACC disabled (OFF condition). In particular, the subject had to assess, on a scale from 0 to 100, the impact of six different factors (i.e. Mental demand, Physical demand, Temporal demand, Performance, Effort, Frustration) and the final result of this questionnaire is a score from 0 to 100 corresponding to the driver's mental workload perception.

3 Results

The analysis of Perception-Reaction Time results, as reported in Fig. 7, showed that during the driving with ACC the mean value of PRT was equal to 3.95 ± 1.92 s, while without ACC it was equal to 2.86 ± 1.02 s. The paired t-test revealed that such a difference was statistically significant (p = 0.023).

The mean Perception-Reaction Time of the drivers with ACC (blue bar) was higher than without ACC (red bar).

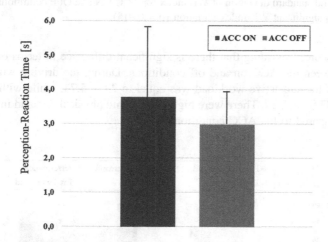

Fig. 7. Mean and standard deviation of PRT for ACC ON and OFF conditions. The paired t-test revealed a significant Perception-Reaction Time variation (p = 0.023). (Color figure online)

In addition, the paired t-test between the WL indexes during the two conditions showed that the WL indexes during the ACC ON driving were slightly higher (p = 0.015) than those during the ACC OFF one (Fig. 8). With ACC the mean value of WL index was equal to 3.22 ± 1.71, whilst without ACC it was equal to 3.04 ± 1.28.

Adaptive Cruise Control caused distraction in the drivers, who exhibited longer PRT times and higher workload. On the contrary, when they knew that they could not rely on the system, they had complete control of the vehicle and they paid more attention to the driving scene, with a consequent Perception-Reaction Time decrease. In response to a braking lead vehicle, drivers started braking later when using ACC. Drivers rate driving with ACC was less effortful when driving without ACC. If PRT is shorter, the driver has a greater probability to be able to stop the vehicle safely with a great improvement in terms of road safety. Figure 9 showed the results in terms of

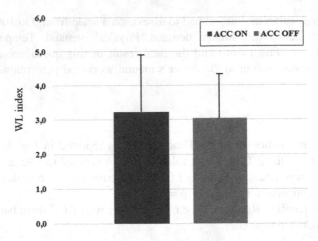

Fig. 8. Mean and standard deviation of WL index for ACC ON and OFF conditions. The paired t-test revealed a significant WL index variation (p = 0.015).

NASA-TLX scores, revealing that there is significant difference in terms of subjective workload between the ACC on and off conditions. During the driving with ACC the mean value of the subjective workload was equal to 29 ± 5.74, while without ACC it was equal to 37.81 ± 2.17. There were higher effort and physical demand in the manual condition compared to the ACC condition.

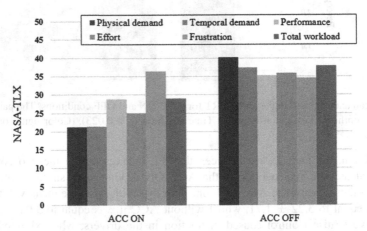

Fig. 9. Mean and standard deviation of NASA-TLX scores for ACC ON and OFF conditions.

The ACC system certainly seemed to fulfil its role as a comfort and convenience device, as it reduced drivers' subjective workload when compared to manual driving. Drivers believed they had maintained a high level of attention during the test dealing with dangerous situations promptly.

This, actually, led to increased distraction and higher Perception – Reaction Times with ACC. Subjects reacted more slowly to a safety-relevant brake light detection task, and responded within an unsafe time when using ACC.

4 Discussion and Conclusions

The Advanced Driver-Assistance Systems (ADAS) positively influence the factors concerning road safety, but they also have effects on the behavior of drivers. In this paper, an experimental test has been carried out in order to evaluate if the use of the Adaptive Cruise Control system could influence the Perception-Reaction Times and mental workload of drivers. The results obtained demonstrated the reliability and effectiveness of the proposed methodology based on human EEG signals, to objectively measure driver's mental workload, and on Mobile Eye-XG device, to evaluate the Perception-Reaction times of drivers. The proposed approach should allow investigating the relationship between human mental behavior, performance and road safety. The results achieved showed that the mean Perception-Reaction Time and the mean mental workload with ACC were higher than without ACC. Results from this study demonstrate that ACC system induced behavioral adaptation in drivers, in terms of changes in workload and hazard detection. Subjects reacted more slowly to a safety-relevant brake task, when using ACC. When the Adaptive Cruise Control was active, it caused distraction in the drivers who exhibited long reaction times. NASA-TLX questionnaire, however, showed there was higher subjective workload in the manual condition compared to the ACC condition. These data are even more significant considering that subjects wore eye tracking glasses and EEG cap, drove an unfamiliar car and knew that their driving behavior was being studied. One may assume that their driving style was more careful than under real-life conditions. From a larger point of view, the present study also demonstrated how such a multimodal evaluation, integrating traditional measures (e.g. car parameters) with innovative methodologies (i.e. neurophysiological measures such as EEG and ET), could provide new and more objective insights. Actually, contrarily to the self-perception, to drive with ACC ON produced higher workload, probably because the drivers were distracted other actions within the car. Therefore, the higher workload could be considered as an indirect effect of ADAS systems, since actually the mere action of driving is perceived as easier by the drivers. However, this results in larger PRT values and therefore in risky behaviors. This preliminary study paves the way to the application of these methodologies to evaluate in real conditions human behavior related to road safety, also considering the recent technological advancements that are making this instrumentation less invasive and easier to use [42, 43]. Further studies are encouraged in order to enlarge experimental sample as well as to treat also other kinds of events as well as ADAS technologies.

Acknowledgments. The authors are grateful to the Unipol Group Spa and, in particular, ALFAEVOLUTION TECHNOLOGY, for the considerable help given in the research study. This work has been also co-financed by the European Commission through the Horizon2020 project H2020-MG-2016 Simulator of behavioural aspects for safer transport, "SimuSafe", (GA n. 723386), and the project "BrainSafeDrive: A Technology to detect Mental States During Drive for improving the Safety of the road" (Italy-Sweden collaboration).

References

1. World Health Organization: Global Status Report on Road Safety 2018
2. Graydon, F.X., et al.: Visual event detection during simulated driving: identifying the neural correlates with functional neuroimaging. Transp. Res. Part F Traffic Psychol. Behav. **7**, 271–286 (2004)
3. Bucchi, A., Sangiorgi, C., Vignali, V.: Traffic psychology and driver behavior. Procedia Soc. Behav. Sci. **53**, 972–979 (2012)
4. Dondi, G., Simone, A., Lantieri, C., Vignali, V.: Bike lane design: the Context Sensitive Approach. Procedia Eng. **21**, 897–906 (2011)
5. Di Flumeri, G., et al.: EEG-based mental workload neurometric to evaluate the impact of different traffic and road conditions in real driving settings. Front. Hum. Neurosci. **12**, 509 (2018)
6. Fan, J., Smith, A.P.: The impact of workload and fatigue on performance. In: Longo, L., Leva, M.C. (eds.) H-WORKLOAD 2017. CCIS, vol. 726, pp. 90–105. Springer, Cham (2017). https://doi.org/10.1007/978-3-319-61061-0_6
7. Salmon, P.M., Young, K.L., Lenné, M.G., Williamson, A., Tomesevic, N., Rudin-Brown, C. M.: To err (on the road) is human? An on-road study of driver errors. In: Proceedings of the Australasian Road Safety Research, Policing and Education Conference, Canberra (2010)
8. Stanton, N.A., Marsden, P.: From fly-by-wire to drive-by-wire: safety implications of automation in vehicles. Saf. Sci. **24**, 35–49 (1996)
9. Chira-Chavala, T., Yoo, S.M.: Potential safety benefits of intelligent cruise control systems. Accid. Anal. Prev. **26**, 135–146 (1994)
10. Rudin-Brown, C.M., Parker, H.A.: Behavioural adaptation to adaptive cruise control (ACC): implications for preventive strategies. Transp. Res. Part F Traffic Psychol. Behav. **7**, 59–76 (2004)
11. Brown, C.M.: The concept of behavioural adaptation: does it occur in response to lane departure warnings? In: International Conference on Traffic and Transport Psychology, pp. 1–14 (2000)
12. Smiley, A.: Behavioural adaptation, safety, and intelligent transportation systems. Transp. Res. Rec. **724**, 47–51 (2000)
13. Fancher, P.S., et al.: Intelligent cruise control field operation test. NHTSA Report No. DOT HS 808 849 (1998)
14. Stanton, N.A., Young, M., McCaulder, B.: Drive-by-wire: the case of driver workload and reclaiming control with adaptive cruise control. Saf. Sci. **27**, 149–159 (1997)
15. Hoedemaeker, M., Kopf, M.: Visual sampling behaviour when driving with adaptive cruise control. In: Proceedings of the Ninth International Conference on Vision in Vehicles, Australia, 19–22 August 2001
16. de Winter, J.C.F., Happee, R., Martens, M.H., Stanton, N.A.: Effects of adaptive cruise control and highly automated driving on workload and situation awareness: a review of the empirical evidence. Transp. Res. Part F Traffic Psychol. Behav. **27**, 196–217 (2014)
17. Pierre, P., et al.: Fatigue, sleepiness, and performance in simulated versus real driving conditions. Sleep **28**, 1511–1516 (2005)

18. Tong, S., Helman, S., Balfe, N., Fowler, C., Delmonte, E., Hutchins, R.: Workload differences between on-road and off-road manoeuvres for motorcyclists. In: Longo, L., Leva, M.C. (eds.) H-WORKLOAD 2017. CCIS, vol. 726, pp. 239–250. Springer, Cham (2017). https://doi.org/10.1007/978-3-319-61061-0_16

19. Aricò, P., et al.: Adaptive automation triggered by EEG-based mental workload index: a passive brain-computer interface application in realistic air traffic control environment. Front. Hum. Neurosci. 10, 539–552 (2016)

20. Prinzel, L.J., Freeman, F.G., Scerbo, M.W., Mikulka, P.J., Pope, A.T.: A closed-loop system for examining psychophysiological measures for adaptive task allocation. International Journal of Aviation Psychology 10, 393–410 (2000)

21. Di Flumeri, G., et al.: Mental workload assessment as taxonomic tool for neuroergonomics. In: Ayaz, H., Dehais, F. (eds.) Neuroergonomics. The Brain at Work and in Everyday Life. Elsevier, Amsterdam (2018)

22. Aricò, P., et al.: Human factors and neurophysiological metrics in air traffic control: a critical review. IEEE Rev. Biomed. Eng. 10, 250–263 (2017)

23. Hogema, J.H., Janssen, W.H., Coemet, M., Soeteman, H.J.: Effects of intelligent cruise control on driving behaviour: a simulator study. In: Proceedings of the 3rd World Congress on Intelligent Transport Systems (1996)

24. Costa, M., Bonetti, L., Bellelli, M., Lantieri, C., Vignali, V., Simone, A.: Reflective tape applied to bicycle frame and conspicuity enhancement at night. Hum. Factors 59, 485–500 (2017)

25. Costa, M., Bonetti, L., Vignali, V., Bichicchi, A., Lantieri, C., Simone, A.: Driver's visual attention to different categories of roadside advertising signs. Appl. Ergon. 78, 127–136 (2019)

26. Costa, M., Simone, A., Vignali, V., Lantieri, C., Bucchi, A., Dondi, G.: Looking behavior for vertical road signs. Transp. Res. Part F Psychol. Behav. 23, 147–155 (2014)

27. Dienes, Z.: Assumptions of subjective measures of unconscious mental states: higher order thoughts and bias. J. Conscious. Stud. 11, 25–45 (2004)

28. Aricò, P., Borghini, G., Di Flumeri, G., Colosimo, A., Pozzi, S., Babiloni, F.: A passive brain–computer interface application for the mental workload assessment on professional air traffic controllers during realistic air traffic control tasks. Prog. Brain Res. 228, 295–328 (2016)

29. Di Flumeri, G., Aricò, P., Borghini, G., Colosimo, A., Babiloni, F.: A new regression-based method for the eye blinks artifacts correction in the EEG signal, without using any EOG channel. In: Proceedings of the Annual International Conference of the IEEE Engineering in Medicine and Biology Society. IEEE Engineering in Medicine and Biology Society (2016)

30. Elul, R.: Gaussian behavior of the electroencephalogram: changes during performance of mental task. Science 164, 328–331 (1969)

31. Klimesch, W.: EEG alpha and theta oscillations reflect cognitive and memory performance: a review and analysis. Brain Res. Rev. 29, 2–3 (1999)

32. Borghini, G., Astolfi, L., Vecchiato, G., Mattia, D., Babiloni, F.: Measuring neurophysiological signals in aircraft pilots and car drivers for the assessment of mental workload, fatigue and drowsiness. Neurosci. Biobehav. Rev. 44, 58–75 (2014)

33. Olson, P.L., Sivak, M.: Perception-response time to unexpected roadway hazards. Hum. Fact. J. Hum. Fact. Ergon. Soc. 28, 91–96 (1986)

34. Li, Y., Chen, Y.: Driver vision based perception-response time prediction and assistance model on mountain highway curve. Int. J. Environ. Res. Public Health 14, 1–31 (2017)

35. Costa, M., et al.: The influence of pedestrian crossings features on driving behavior and road safety. In: Proceedings of the AIIT International Congress on Transport Infrastructure and Systems, pp. 741–746 (2017)

36. Costa, M., Bichicchi, A., Nese, M., Lantieri, C., Vignali, V., Simone, A.: T-junction priority scheme and road user's yielding behavior. Transp. Res. Part F Traffic Psychol. Behav. **60**, 770–782 (2019)
37. Costa, M., Bonetti, L., Vignali, V., Lantieri, C., Simone, A.: The role of peripheral vision in vertical road sign identification and discrimination. Transp. Res. Part F Traffic Psychol. Behav. **61**, 1619–1634 (2018)
38. Costa, M., Simone, A., Vignali, V., Lantieri, C., Palena, N.: Fixation distance and fixation duration to vertical road signs. Appl. Ergon. **69**, 48–57 (2018)
39. Lantieri, C., et al.: Gateway design assessment in the transition from high to low speed areas. Transp. Res. Part F Traffic Psychol. Behav. **34**, 41–53 (2015)
40. Vignali, V., Bichicchi, A., Simone, A., Lantieri, C., Dondi, G., Costa, M.: Road sign vision and driver behaviour in work zones. Transp. Res. Part F Traffic Psychol. Behav. **60**, 474–484 (2019)
41. Vignali, V., et al.: Effects of median refuge island and flashing vertical sign on conspicuity and safety of unsignalized crosswalks. Transp. Res. Part F Traffic Psychol. Behav. **60**, 427–439 (2019)
42. Aricò, P., Borghini, G., Di Flumeri, G., Sciaraffa, N., Babiloni, F.: Passive BCI beyond the lab: current trends and future directions. Physiol. Meas. **39**(8), 08TR02 (2018)
43. Di Flumeri, G., Aricò, P., Borghini, G., Sciaraffa, N., Di Florio, A., Babiloni, F.: The dry revolution: evaluation of three different EEG dry electrode types in terms of signal spectral features, mental states classification and usability. Sensors **19**(6), 1365 (2019)

Author Index

Printed in the United States
By Bookmasters